Mitigation and Financing of Seismic Risks: Turkish and International Perspectives

NATO Science Series

A Series presenting the results of scientific meetings supported under the NATO Science Programme.

The Series is published by IOS Press, Amsterdam, and Kluwer Academic Publishers in conjunction with the NATO Scientific Affairs Division

Sub-Series

I. Life and Behavioural Sciences	IOS Press
II. Mathematics, Physics and Chemistry	Kluwer Academic Publishers
III. Computer and Systems Science	IOS Press
IV. Earth and Environmental Sciences	Kluwer Academic Publishers

The NATO Science Series continues the series of books published formerly as the NATO ASI Series.

The NATO Science Programme offers support for collaboration in civil science between scientists of countries of the Euro-Atlantic Partnership Council. The types of scientific meeting generally supported are "Advanced Study Institutes" and "Advanced Research Workshops", and the NATO Science Series collects together the results of these meetings. The meetings are co-organized bij scientists from NATO countries and scientists from NATO's Partner countries – countries of the CIS and Central and Eastern Europe.

Advanced Study Institutes are high-level tutorial courses offering in-depth study of latest advances in a field.
Advanced Research Workshops are expert meetings aimed at critical assessment of a field, and identification of directions for future action.

As a consequence of the restructuring of the NATO Science Programme in 1999, the NATO Science Series was re-organized to the four sub-series noted above. Please consult the following web sites for information on previous volumes published in the Series.

http://www.nato.int/science
http://www.wkap.nl
http://www.iospress.nl
http://www.wtv-books.de/nato_pco.htm

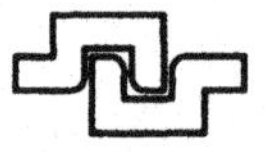

Series IV: Earth and Environmental Series – Vol. 3

Mitigation and Financing of Seismic Risks:
Turkish and International Perspectives

edited by

Paul R. Kleindorfer
Risk Management and Decision Processes Center,
The Wharton School,
University of Pennsylvania,
Philadelphia, PA, U.S.A.

and

Murat R. Sertel
Foundation for Economic Design & Department of Economics,
Boğaziçi University,
Istanbul, Turkey

Kluwer Academic Publishers

Dordrecht / Boston / London

Published in cooperation with NATO Scientific Affairs Division

Proceedings of the NATO Advanced Research Workshop on
Mitigation and Financing of Earthquake Risks in Turkey
Ankara, Istanbul, Turkey
22–24 June, 2000

A C.I.P. Catalogue record for this book is available from the Library of Congress.

ISBN 0-7923-7098-8 (HB)
ISBN 0-7923-7099-6 (PB)

Published by Kluwer Academic Publishers,
P.O. Box 17, 3300 AA Dordrecht, The Netherlands.

Sold and distributed in North, Central and South America
by Kluwer Academic Publishers,
101 Philip Drive, Norwell, MA 02061, U.S.A.

In all other countries, sold and distributed
by Kluwer Academic Publishers,
P.O. Box 322, 3300 AH Dordrecht, The Netherlands.

Printed on acid-free paper

Printed in the Netherlands.

To Ahmet Sertel (1915-2000),
whose vision and energy for Turkey
and ourselves
will always be with us.

CONTENTS

Recent Policy Developments in Turkey

Insurance and Risk Transfer Mechanisms

AUTHORS

Murat Balamir, The Scientific and Technical Research Council of Turkey (TÜBITAK) and Middle East Technical University, Ankara, Turkey

Fouad Bendimerad, Risk Management Solutions, Inc., Menlo Park, California, USA

M. Hasan Boduroğlu, Istanbul Technical University, Istanbul, Turkey

Ferruccio Ferrigni, University of Naples "Federico II", and International Institute Stop Disasters, Naples, Italy

Polat Gülkan, Middle East Technical University, Ankara, Turkey

Paul R. Kleindorfer, University of Pennsylvania, Philadelphia, USA

Howard Kunreuther, University of Pennsylvania, Philadelphia, USA

Martin Nell, Johann Wolfgang Goethe-University, Frankfurt/Main, Germany

Ayse Öncüler, INSEAD, Fontainebleau, France

Ioannis S. Papadakis, University of British Columbia, Vancouver, B.C. Canada

Joanna Papoulia, Institute of Oceanography, National and Center for Marine Research, Athens, Greece

Andreas Richter, Hamburg University, Hamburg, Germany

Horea Sandi, The National Building Research Institute, Bucharest, Romania

Rakesh Sarin, University of California Los Angeles, USA

Murat R. Sertel, Center for Economic Design, Boğaziçi University, Istanbul, Turkey

Tuğrul Tankut, The Scientific and Technical Research Council of Turkey (TÜBITAK) and Middle East Technical University, Ankara, Turkey

Barbaros Yalçin, MilleRe Reinsurance Company, Istanbul, Turkey

William Ziemba, University of British Columbia, Vancouver, B.C. Canada

PREFACE AND ACKNOWLEDGEMENTS

The magnitude 7.4 earthquake on the morning of August 16, 1999 in the region of the Bay of Izmit in Turkey was a clarion call to review and reformulate Turkey's policies towards earthquake hazards. With over 19,000 fatalities, and hundreds of thousands of displaced individuals and businesses, this event caused huge economic and social disruptions in a key economic region of Turkey. One clear result of this event was recognition by government, industry and the public that Turkey could not simply go on as it had in the past. Changes in financial coverage for economic losses from earthquakes, better enforcement procedures for building standards, better business contingency planning for supply disruptions, and well developed emergency response were demanded from all sides. This groundswell gave rise to a radical redesign of the economic and engineering institutions in Turkey responsible for mitigating and financing the risks of seismic activity. But Turkey has not been alone in its recognition of the necessity of radical change in its policies toward natural disasters. Other countries in the Mediterranean region have been pushing similar reforms. Major emerging economies such as India and China and most of Latin America have also joined in the search for new public-private partnerships to promote better risk quantification and ultimately better means of coping with the human and economic costs of natural catastrophes. The NATO Advanced Research Workshop, which took place in Istanbul from June 22-24, 2000, and of which this volume is the Proceedings, was thus a timely meeting, given the strong recognition of the need for action in this vital area from many quarters.

This Proceedings volume contains only the formal papers prepared for the Workshop. But the vitality of the discussion was considerable enhanced by many other contributions that we are pleased to recognize here. Foremost, we were grateful to Özgün Ökmen, from the Office of the Prime Minister of Turkey and various representatives of the General Directorate of Emergency Management in Turkey who accompanied Mr. Ökmen. We also thank Mrs. Serap Oğuz Gönülal, from the General Directorate of Insurance in the Prime Ministry for her contributions to the Workshop. A number of others contributed presentations or leadership roles in the Workshop, including the following individuals to whom we express our gratitude: Attila Ansal, Nuray Aydınoğlu, Mustafa Erdik, Uğur Ersoy, Emil-Sever Georgescu, and Günay Özmen.

As co-directors of the Workshop, we owe a debt of gratitude to a number of individuals and organizations, without whose assistance this Workshop could not have taken place. First, we would like to recognize The Scientific and Technical Research Council of Turkey (TÜBITAK) and its president, Professor Namık Kemal Pak, whose early support and hosting of organizing meetings for the Workshop brought focus and energy to our deliberations. Second, we wish to recognize the members of our Organizing Committee, Fouad Bendimerad, Semih Koray, Tuğrul Tankut, and Seha Tiniç who assisted us in crafting the vision and in executing the detailed planning for the Workshop. Third, we were very grateful for the support and advice of Alain Jubier of the NATO Science Programme in Brussels. The Wharton Project on Managing Catastrophic Risks also provided some support for this Workshop, which is gratefully acknowledged. We also wish to recognize the care and dedication of Jeremy Guenter of the Center for Research in Regulated Industries, Rutgers University, for his editorial assistance.

Finally, this Workshop would not have been possible without the support of our own Centers, the Wharton Center for Risk Management and Decision Processes at the University of Pennsylvania and the Center for Economic Design at Boğazaçi University.

Paul R. Kleindorfer, Philadelphia
Murat R. Sertel, Istanbul
April, 2001

INTRODUCTION

1

Introduction and Overview of the Proceedings

Paul R. Kleindorfer
The Wharton School, University of Pennsylvania

Murat R. Sertel
Center for Economic Design, Boğaziçi University

In this introduction to the Proceedings of the NATO Workshop on Mitigation and Financing of Earthquake Risks, we wish to draw attention to some of the key concerns that any program directed at reducing the risks or the costs of seismic events will have to confront. We then briefly review the contributions of the papers presented at the Workshop and some of the key findings of the Workshop.

The key problem in earthquake mitigation and financing is to focus attention on decisions that will affect ultimate outcomes at the time when these decisions are made and not just in cleaning up the results of bad prior decisions after the fact. From this perspective, the key issue is aligning incentives for property and business owners to undertake appropriate risk mitigation before disasters occur. In this regard, insurance, whether privately or publicly provided, and its sister disciplines of risk management and loss reduction, can play a vital role as a lever for efficiency. Insurance not only pays for some of the economic costs of recovery from such events. It also can provide valuable information on the risks faced by particular areas or types of structures vulnerable to such damage, and thus can signal to the owners of such structures the attendant risks they face. Coupled with established actuarial and engineering practices for risk assessment, such signals can provide appropriate pressures for risk mitigation in structures, in location choices and in emergency response. At the very least, these signals can lead to a greater awareness of the existence of major seismic risks.

While the ideas of risk quantification and loss reduction, including insurance, are not new, the devil is in the details, and these are likely to be quite specific to the country or jurisdiction in question. The first issue, of course, is the difficult and country-specific problem of establishing an

P.R. Kleindorfer and M.R. Sertel (eds.), Mitigation and Financing of Seismic Risks, 3–9.

accurate inventory of various property types by location and establishing vulnerability characteristics of each property type via structural engineering. Other specific problems associated with risk assessment and mitigation include the following:

1. Many buildings in countries like Turkey are quite old and enjoy considerable cultural value in themselves and, as an externality, in association with the characteristics of other buildings in their immediate neighborhood. This means that the process of determining and implementing appropriate standards for retrofitting must engage not only engineering professionals but the local community.
2. There are a number of peculiarities in the insurance industry in every country, including how well developed and capitalized the industry is. This is so whether catastrophe insurance is privately funded, as it is in some countries, publicly funded, as it is in many other countries, or a mixture of the two, as it is in Turkey. Thus, determining the appropriate fit between economic theory and existing insurance and risk-bearing structures is an important matter that will require the cooperation between public and private parties associated with providing and regulating insurance.
3. There may be significant problems in convincing property owners with tight budget constraints to pay for insurance or mitigation measures. Subsidizing insurance or seismic retrofitting of buildings is one approach, but such subsidies lead naturally to moral hazard and may undermine the very incentives one is attempting to promote through an insurance-based program.
4. The political process at the municipal and regional level may itself contribute to the problems of assuring preparedness for seismic events, as public officials have on occasion bypassed or not enforced safe building standards, even for new construction. Clearly, whatever economic instruments are proposed must also be compatible with existing political institutions.

The papers in this Workshop address these questions and summarize the current state of related research. Before reviewing these, we note an important issue concerning insurance in financing catastrophe risks of special relevance in the catastrophe insurance area. For many emerging economies, including Turkey, the private insurance system is not sufficiently well developed or capitalized to allow it to cover the huge potential losses associated with natural hazards. The large potential losses from natural hazards, including floods, hurricanes and earthquakes, arise because of the correlated loss distributions characteristic of natural hazards, giving rise to long-tailed risks that complicate the usual pooling efficiencies of private insurance. This has given rise to a public-private partnership in which the Government plays a major role in assessing and bearing the risks from natural hazards, at least until the private insurance system is able to "crowd out" the Government from this role. In such countries, private insurance still plays an important role in the commercial and industrial sector, but not in the homeowner sector. Thus, in all emerging economies, it is to be expected that the public sector will play an important role in financing catastrophe risks.

Turning now to the contributions of the papers in this volume, we begin with the most important scientific foundation for effective management of natural disasters, developing the necessary infrastructure for assessing the risks a country faces from these natural hazards. As described in the Bendimerad, Papoulia and Sandi papers, a number of scientific and engineering disciplines provide input to risk quantification and risk modeling to estimate losses and to provide necessary inputs for risk-bearing programs, such as national insurance. The now accepted process for doing this estimation is to use large-scale simulation models for the hazard in question. For example, for seismic hazards in the Istanbul region, a list of several thousand representative seismic events would be determined (differentiated by location and magnitude) and, for each of these events, the losses resulting from the event for a set of structures would be individually evaluated by simulating the likely damages for each structure and each event. Thereafter these event-specific loss distributions would be further weighted by the estimated recurrence times for each of the events, to obtain finally the (estimated) loss distribution for the given set of structures during a specified time period, e.g., a year. By coupling this with emergency response requirements, both public and private parties engaged in mitigation or financing of seismic risks can determine before the fact the magnitude of the potential losses they face and what they can do to reduce these.

The above narrative applies equally well to developed and emerging economies, but some of the details are more challenging in emerging economies. For example, for earthquake exposures, the precise location of an insured structure is extremely important for an accurate simulation of any of the events in the hazard event table driving these simulations. Very often, however, accurate building inventories, including specific information on the type of each insured structure, are not available. Also, knowledge of the vulnerabilities of specific structures may be lacking (e.g., because of lack of standardization of building design or uneven enforcement of building codes). Overcoming these problems requires a concerted effort organized along several dimensions. We discuss the further contributions of this volume to this question under two headings, the demand side and the supply side.

1. DEMAND SIDE: PROMOTING EFFICIENT MITIGATION AND FINANCING OF SEISMIC RISKS

If homeowners and businesses acted according to the dictates of economic rationality, the matter would be relatively straightforward. Risk-averse consumers would evaluate the relative merits of insurance and other protective measures, such as structural mitigation, and would adopt cost-effective choices where available. Public or private insurers would supply the basic signals through prices of insurance derived from catastrophe modeling of the expected losses associated with various locations, building types and mitigation measures undertaken. The result would not be an absence of risk, of course, but rather an alignment of the costs of risk with other opportunity costs associated with such decisions.

The real problem, however, is made more difficult by the apparent difficulty that homeowners and businesses have in evaluating insurance and mitigation decisions. As noted in the Kunreuther and Öncüler papers, people appear to be afflicted by a variety of imperfections in evaluating such decisions. Thus, well-enforced building codes and zoning laws are a critically important element of natural hazard mitigation. So are retrofitting standards and knowledge, as the Sarin paper points out in the context of Los Angeles. The papers by Balamir, Ferrigni, Gülkan and Tankut discuss how such building codes have been designed in Italy, Turkey and Romania, together with the associated regulations, laws and spatial planning procedures, including zoning, that are required to implement these codes and to accomplish the end of assuring seismically fit buildings.

One approach to promoting property owner acceptance of retrofitting or other mitigation measures is to tie insurance premiums, lower deductible levels (and other desired policy features) to the implementation of approved mitigation measures. Another approach is to use community awareness and emergency response programs to inform property owners of the risks they face and what they can do about them, especially if these risks can be life threatening or if they can be severely disruptive of economic well being. The importance of preparedness and the organization of emergency response is well illustrated in the paper by Papadakis and Ziemba in the context of Taiwan's dramatic recovery in its business sector following the major earthquake there in September, 1999.

Certainly one important feature of any efficient economic design for loss reduction and insurance will remain the precision of insurance pricing signals. If property owners see no difference in insurance rates as a result of their actions or the vulnerability their property to natural hazards, the incentives for them to undertake changes in their structure or their insurance decisions will be obviously be weakened. For the same obvious reason, if the government provides highly subsidized ex post disaster relief roughly comparable to what can be obtained by purchasing insurance ex ante, the demand for insurance will quite properly go to zero, and so would any incentives for other ex ante protective activity. It should be plain that while property owners may have difficulty with some of the complexities of catastrophe insurance, they will have no difficulty seeing through government charity.

Thus, on the demand side, establishing a sustainable program for mitigating and financing seismic risks requires the presence of a number of simultaneous factors, including the necessary political will to avoid a post-disaster rescue culture, community discipline to establish and enforce proper building codes, and a broad public information program to inform the public on the risks they face. These ingredients are difficult to establish in the best of times, and it is not surprising that catastrophe risks have continued to pose major problems for governments and private insurers around the world, even in developed, market-oriented economies. In emerging economies, where the starting point is a much less viable private insurance market, these problems have led to various state-supported insurance solutions, often coupled with compulsory insurance and a national or regional pool, as the case in France, New Zealand, Spain and Turkey. As the Boduroğlu and Yalçin papers discuss, these considerations, and the desire to provide adequate funding for the new Turkish Catastrophe Insurance Fund, led to compulsory insurance in Turkey. This mandatory approach overcomes many of the demand side problems noted above, but at the cost of dictating at least minimum levels of insurance, even when these are not desired by an informed property owner.

2. SUPPLY SIDE: ASSURING A SUPPLY OF RISK-BEARING CAPITAL AND TRAINED ENGINEERS

Supply side failures are not hard to find as well. Many property and casualty insurers were completely surprised by the losses they suffered in major past events (such as hurricane Andrew (1992) in Florida or the Northridge earthquake (1994) in California, as noted in the Kunreuther paper). Indeed, the low levels of surplus that many of these insurers had prior to major events suggested something between blissful ignorance and irresponsible gambling behavior. These problems arise in part because of the complexity of understanding and quantifying the risks associated with natural catastrophes. Given the low probability of these events and the short historical record we have to validate structural damage models and the recurrence frequency of natural hazards, it is not surprising that catastrophe models provide estimates with large uncertainty bounds. These uncertainties may be further exacerbated by poor data on building locations, types and on the surrounding geological conditions or topography. These uncertainties considerably complicate the underlying actuarial mathematics associated with risk-based pricing.

A further difficulty in financing catastrophe risks is obtaining reinsurance, either for private insurers or national pools. The use of new capital market instruments for emerging economies, as a supplement to traditional reinsurance, is analyzed in detail in the Nell and Richter paper and, in connection with earthquake risks in Turkey, in the Yalçin paper. As they note, the key challenge for economic design in this regard is the design and pricing of reinsurance contracts and capital market instruments sufficiently attractive to investors to provide the capital base to bear the risks of hazards. Doing so will require as a precondition a respected scientific infrastructure for quantifying the risks that underlie these hazards.

3. SOME POLICY CONCLUSIONS

A key insight derived from the Turkish and French experience is the use of existing insurance industry as infrastructure for distributing policies, with ultimate risk bearing undertaken by the Government. The first key challenge for any effective policy dealing with natural hazards must be the development of an adequate infrastructure for risk quantification, whether or not the Government is involved. Workshop participants pointed to the virtuous cycle of accurate risk assessments, risk-based premiums for (public or private) insurance, awareness by property owners of the risks they face, cost-effective mitigation, valued service provided by insurance and emergency response organizations in the event of disasters, and increased public trust in the institutions associated with natural disaster mitigation and financing. This virtuous cycle is unfortunately easy to break, at any step, but it begins with a solid science and engineering base for risk quantification.

A key question raised in the Workshop about the current Turkish situation is whether there is (or will be) sufficient political will to deny benefits to those property owners who do not purchase insurance, as required by the new compulsory insurance law. It will take some time to establish a social norm to buy the insurance as a part of social responsibilities of citizenship. It will take even longer to establish the value of such insurance as a real service to policyholders, a friend in need in the post disaster world. One of the ways suggested for promoting public trust in programs like Turkey's compulsory insurance program is to use some of the funds collected under the program to promote visible mitigation projects, especially for critical structures such as schools and hospitals.

Several proposals arose in the Workshop for the development of a voluntary, private insurance system. This will be especially important for an emerging economy, like Turkey, that is increasingly integrated with the global economy. Disruptions in economic activity through natural disasters, if not covered by appropriate insurance, will be damage not just properties but on-going supply chains connecting the economy to its international trading partners. Fortunately, establishing viable insurance markets for commercial and industrial properties, and for business disruption, would not seem to be a major problem for fast growing emerging economies as long as there is freedom by private insurers to charge appropriate risk-based rates. Buildings and businesses that are "seismically fit" would then enjoy very significant benefits in reduced premiums.

These are important issues in seeing the development of strong base of engineering knowledge in any economy. The Workshop underlined many of the difficulties in applying this knowledge to mitigate seismic risks, however. For example, even with the technical capacity to retrofit, there will be significant problems associated in large cities with joint ownership of multi-story buildings. Nonetheless, a key factor for any local, regional or national program of risk mitigation will be the rational use of engineering knowledge in supporting cost-effective risk mitigation. Any approach to the problem must be considered from a long-range perspective since the problem is a long-term one. Moreover, the driving force in most emerging economies for public acceptance of building codes, retrofitting and other structural mitigation programs remains saving lives, not avoiding economic damages.

Key stakeholders involved in implementing any coherent program for disaster preparedness and prevention are the community and the technical professions, from actuaries to engineers. For complex and historical cities like Athens, Bucharest, Istanbul or Rome, these stakeholders together can craft a reasoned approach to cope with natural hazards. The good news is that a number of working and workable models for public-private partnership to save lives and mitigate economic disruptions from natural hazards are being implemented around the world. This Workshop provided an important marker of the progress that has been made to date in developing these models, and an opportunity to compare the approaches of different countries in addressing this issue going forward.

ESTABLISHING THE INFRASTRUCTURE FOR RISK QUANTIFICATION

2

Modeling and Quantification of Earthquake Risk:

Application to Emerging Economies

Fouad Bendimerad
RET-International
San Jose, California USA

INTRODUCTION

Earthquakes and other natural hazards can create disasters of uncontrollable magnitudes when they hit large metropolitan areas. Losses from catastrophes are increasing due to higher concentrations of people and value in catastrophe-prone areas, and vulnerability of infrastructure. From 1994 to 1999, several earthquakes caused financial losses in excess of US$10 billion, including Northridge Earthquake, Los Angeles, 1994, Great Hanshin Earthquake, Kobe, 1995, the Kocaeli Earthquake, Turkey, 1999, and the Chi-Chi Earthquake, Taiwan 1999. According to a World Bank publication, over 100,000 people died as a result of natural and technological disasters in 1999, with economic costs exceeding US$65 billion worldwide (Cameron, 2000). In developing countries, most of the economic loss falls under the burden of the Government, which has to divert precious development and public services funds into the recovery and reconstruction process. Business owners and homeowners, who are victims of disasters, often rely on disaster assistance help from government to rebuild their source of income and houses because they generally do not carry earthquake insurance. Insurance penetration in emerging economies is a fraction of that of developed countries, and plays a minimum role in protecting the economies and welfare of developing countries. The low level of insurance is indicative of an inadequate understanding of the role of insurance in catastrophe loss protection in emerging economies.

P.R. Kleindorfer and M.R. Sertel (eds.), Mitigation and Financing of Seismic Risks, 13–39.

Catastrophes such as Northridge together with projections of even higher losses in the future pushed the global insurance industry in the US and other developed nations, to adopt catastrophe models to manage their catastrophe exposure. Catastrophe models represent a more rational way to measure catastrophe loss potential than other techniques because they do not solely rely on past losses and they incorporate the physical sciences controlling the natural hazard behavior and the damage patterns to exposed property. The practicality of analyzing large portfolios of insured property was made possible by advances in computing technology that allows manipulation of large amount of data together with the capability of performing complex analytical algorithms.

Disaster management is another area of application of catastrophe models. Governments, development agencies, humanitarian agencies, private and public entities have become increasingly involved in catastrophe risk management. There is great concern over population safety, protection of property and safeguard of investments. Recently, governments in the US, Japan, Australia and other developed nations have also sponsored the development of catastrophe models geared towards disaster planning. In particular, in the United States, the Federal Emergency Management Agency (FEMA) made the development of disaster management technology a key item in its agenda to mitigate the effects of disasters. FEMA funded the development of the HAZUS technology, a standardized earthquake loss estimation methodology, which is used by government and non-government institutions to estimate potential losses from earthquakes and to establish local mitigation programs (NIBS 1997, 1999).

The incorporation of catastrophe models in disaster management has enlarged the field of application of these models, fueling interest from a large constituency of users. However, there is a gap in the use of technology between developed economies and emerging economies. In developed economies, the use of catastrophe models has become standard practice in quantifying, underwriting and transferring risk due to earthquakes and other disasters. Catastrophe models are also used to prepare communities and institutions for disasters. In contrast, emerging economies do not generally use modeling technology to manage their catastrophe exposure. Typically, government regulators set up tariffs; underwriting practices rely on individual initiative; and the global reinsurance market controls risk transfer conditions. Government focuses its role on response and rebuilding efforts instead of pro-active mitigation. Catastrophe modeling technology could drive progress in emerging economies' ability to manage the impacts of earthquakes and other disasters.

1. BACKGROUND

Damage and loss estimation techniques were first used for disaster planning. In the late 1970's, effects of potential catastrophes were estimated from scenario analyses, by which a postulated event was assumed and its potential losses calculated for a pre-determined exposure. Lack of access to computing technology and major gaps of knowledge have caused these studies to be arduous and time consuming. A loss estimation study for a major metropolitan area could take months to collect the underlying data and would require the participation of experts from several fields, thus limiting the practical application of this technique. Despite their complexity, loss estimation studies have proven to be a very useful tool for developing emergency preparedness plans and for promoting seismic risk mitigation (National Research Council, 1989; FEMA 1994).

Private vendors started developing desktop software packages aimed at estimating insurance catastrophe losses in the late 80's. These software packages, made possible by the information technology revolution, were able to represent an actual portfolio of insured risks, and calculate the losses by geographical territory as well as the aggregated losses to the portfolio for a series of simulated events. The insurance industry was not adequately equipped to understand the inner working of the models and remained skeptical about the benefit of their application and the validity of their outputs. For an industry that looked at the future through the experience of the past, the inability of the models to consistently match past experience represented a serious handicap. More importantly, prior to suffering large catastrophe losses, insurers did not necessarily see a pressing need for catastrophe models because they felt fairly confident that they could fund losses, particularly with the availability of relatively inexpensive reinsurance capacity. Reinsurers rarely experienced significant catastrophe losses, and were comfortable that diversification across a large number of cedants and geographical areas would limit their maximum losses. Events such as Northridge, however, proved that the projected losses were grossly understated, and sent insurers and reinsurers in a quest for models that could help them manage their catastrophe exposure more rationally.

Within the last few years, catastrophe models gained a remarkably high profile in the property and casualty insurance industry. Today's leading catastrophe models are capable to computing probable loss distributions for individual risks, multi-location policies and certificates, treaties, company portfolio and even the entire insurance industry based on the most recent scientific and technical research. As opposed to estimating the impact of deductible and retention on a statistical basis, these models actually compute the expected losses to individual layers and participants based on the coverage structure. In addition, they are able to estimate uncertainty and are accessible to a whole spectrum of users from decision-makers to staff technicians.

Both the application of technology and the capabilities of the technology itself are continuing to improve rapidly. Today, global reinsurers keep increasing their capability to model risk around the globe and their internal understanding of the risk profile of many emerging economies. Such knowledge provides them with a clear advantage in establishing market condition. Domestic insurers and regulators in emerging economies lack pertinent knowledge of their risk profiles and play a minimum role in establishing reinsurance rates for their own portfolios. They have to abide by the rules of the global reinsurance market in terms of cost and capacity of reinsurance. As a case in point, at the time of the renewals of the 2001 catastrophe treaties, reinsurance rates in the Eastern Mediterranean countries including Turkey and Greece have more than double as a consequence of the August 17, 1999 Kocaeli Earthquake and the September 7, 1999 Athens Earthquake. The lack of use of catastrophe models to evaluate the earthquake risk to emerging economies precludes an objective comparison between the rates and the actual risk-on-line.

The lack of pertinent risk parameters also limits the influence of domestic insurers in reducing catastrophe exposure and, thus protecting the economy. The absence of market dynamics inhibits the growth of insurance penetration and hinders market competition and consumer confidence. As a result, catastrophe insurance in emerging economies in highly inefficient and cannot effectively sustain the economic activity of these countries, shifting the burden of catastrophes on individuals and government instead. The use of catastrophe models would contribute to leveling the field between demand and supply, would introduce rational underwriting and rating methods, would increase efficiency in the process of financing and transferring catastrophe risk, and would stimulate the insurance industry in emerging markets.

2. GENERAL DESCRIPTION OF MODEL

Catastrophe models used for risk transfer[1] purposes are proprietary. Providing a description that applies to all models is therefore difficult

---

[1] From this point on, the paper will focus on the risk-transfer applications of earthquake catastrophe models

because modeling companies do not fully disclosed the details of the methodologies and architecture of their models. This section provides a generic description of the architecture of a catastrophe model with emphasize on discussing the underlying scientific and engineering methodologies behind the models. For completeness and wherever appropriate, options for modeling are discussed to provide a comprehensive view of the different available approaches. This description is not meant to represent any particular commercial model.

The architecture of a catastrophe model typically consists of four logically inter-connected modules: The Exposure Module, the Hazard Module, the Vulnerability Module and the Financial Module. Data and results of calculations done in one module are passed on to the next module to produce various loss components (typically loss to the insured, loss to the insurer and loss to the reinsurer). The calculations are done for each location, then aggregated over all locations to produce the components of loss to the portfolio that is analyzed. A schematic of a model is shown in Figure 1. A summary description of each module follows:

2.1 Exposure Module

This module captures the input data that defines the portfolio to be analyzed. It consists of the following parameters:

- Location of the properties in the portfolio
- Description of the properties (i.e., construction type, age, number of stories)
- Occupancy (e.g., residential, commercial, industrial, etc.)
- Value of properties at risk
- Insurance contract information (i.e., policy details prescribing type and amount of coverage, deductibles and limits)
- Reinsurance information (this could include a number of reinsurance structures that are applicable to the site, to an account composed of several sites, or to the portfolio)
- Additional information about the properties at risk such as detailed construction characteristics (e.g., soft story, cladding type, etc.)

The exposure information is organized in relational databases that communicate with a geographical information system (GIS) which maps the information about each of the locations. A geocoding engine tracks the information on each location and assigns it longitude-latitude coordinates. In addition, the geocoding engine has the important function of interpreting the geographical resolution of the exposure information.

There are two distinct dimensions of resolution. First, the resolution in the exposure information, which relates to how well insurers and reinsurers capture data on exposure; and second the resolution in the analysis which relates to the ability of the model to correlate the calculations on hazard and vulnerability to the exposure characteristics of a particular location. These two dimensions of resolution vary greatly. In term of analysis resolution, the most sophisticated models are able to take street level exposure information and calculate hazard at the coordinates of each property; they will also attempt to represent the site characteristics and the damageability of each property in the portfolio. This is generally referred to as 'street-level' resolution. Note that this level of resolution would also require information on soil conditions for the location of each property. Street level resolution is only typically available for a few large insured commercial properties. Even in such cases, detailed construction information as well as soil information may not be available. Nonetheless, some models support analysis at street level resolution, at least in terms of hazard calculations.

Typically, insurers aggregate their exposure data at Postal Code level (or a geographical configuration similar to a postal code). In this case, the geocoding engine will correlate the information to a centroid of the Postal Code. The position of the centroid is calculated either as the position of the geometrical centroid, or some weighted position where the weight depends on information such as population distribution or insurance exposure distribution. Most models are designed to work at Postal Code level resolution or equivalent. Exposure information on portfolio is aggregated at the centroid of the Postal Code and hazard is calculated at the coordinates of the centroid and assumed to be uniform over the whole area of the Postal Code. Soil data, liquefaction and landslide potential are also assigned to the entire area of the Postal Code. In some cases, exposure information is at even lower resolution, such as county-level or even Cresta[2] level. In this case, the models capture the information and then redistribute the exposure at a higher level of resolution, such as Postal Code according to internal inventory mapping databases. The analysis is typically performed at Postal Code level or an equivalent geographical grid.

---

[2] Crestas are large geographical territories conventionally used by the insurance industry worldwide to represent accumulations.

Figure 1: Schematic of an Earthquake Risk Evaluation Model

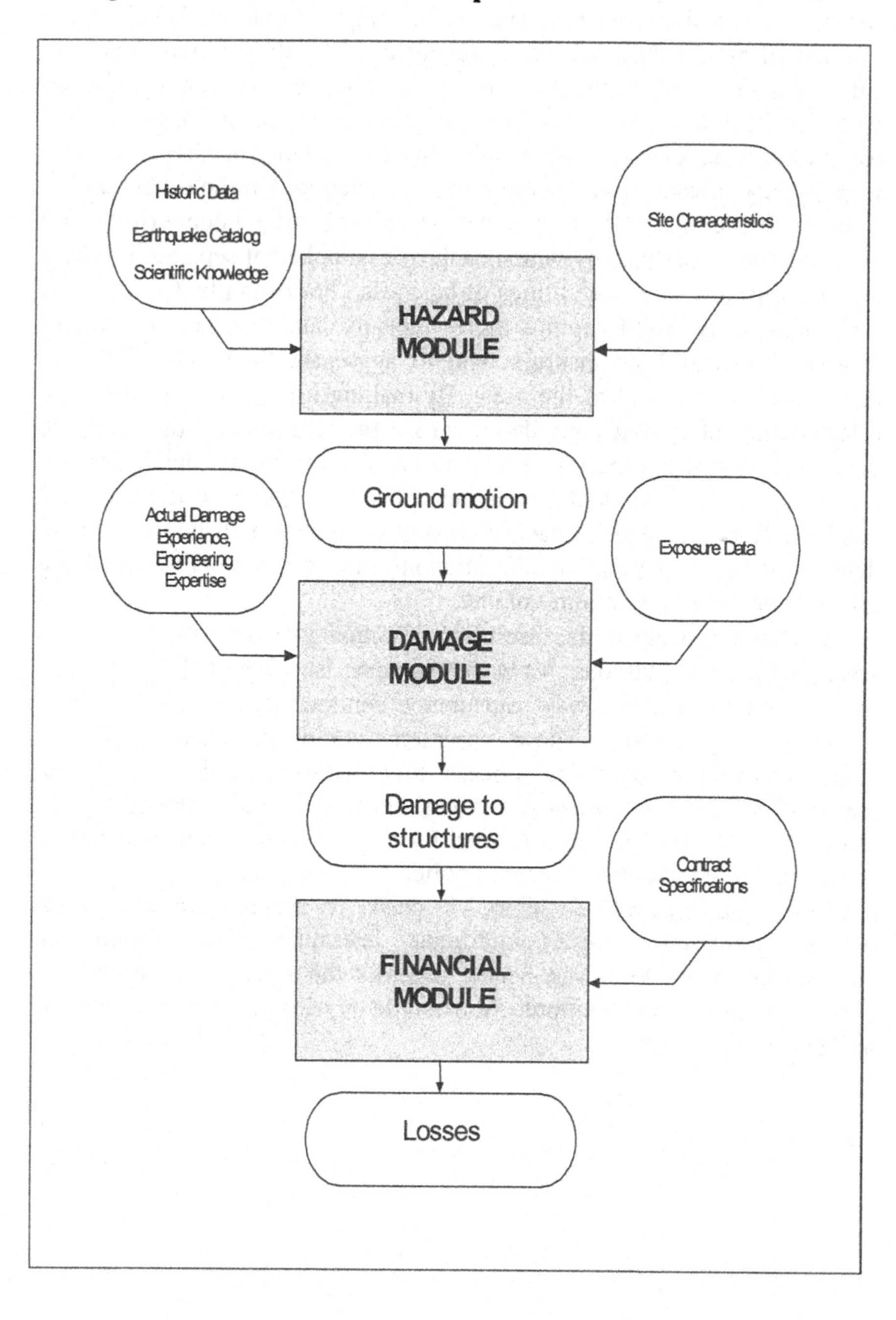

Typically, before, performing an analysis, the data is audited for completeness and consistency, and checked against industry benchmarks. The model architecture is also designed to deal with different levels of data resolution. The application for an insurer and a reinsurer could be quite different due to the nature of their businesses. A reinsurer typically receives submissions at a low resolution (e.g., County or Cresta levels[3]) with few specifications on construction information. On the other hand, a primary insurer may use the model to enforce strict underwriting and accumulation guidelines by which each risk is looked at separately. There are also differences between lines of business. For example, Commercial Line insurers typically capture more relevant data about each property whereas Personal Line insurers tend to aggregate their data. Models reflect the environment of the user. By making use of internal inference rules and inventory databases they improve the resolution of the exposure data, and complete missing information. The inventory databases are proprietary to each modeling company. They are based on industry and portfolio data collected by each company combined with public domain data such as - population distribution, and commercially available information such as premium volume.

Another parameter in the resolution of a model pertains to its ability to adequately model various ways reinsurance is contracted. There are several variations in the way reinsurance contracts operate (Kiln, 1981; Carter et. al., 2000). These variations could introduce significant complexities in the modeling process. First, a model needs to recognize the nature of a reinsurance contract (e.g., facultative, quota-share, surplus share, etc.). Second it needs to identify the levels at which reinsurance contracts and catastrophe treaties applies (i.e., location, policy, account, portfolio, and corporate). Third, it needs to represent inuring laws between contracts. These conditions determine how a reinsurer participates in the loss, and hence, controls the accuracy of each loss component of the model output. An example of reinsurance rules is shown in Table 1.

---

3 Crestas are geographical territories conventionally used by both insurers and reinsurers to delineate accumulations and exposures.

Table 1: Example of Reinsurance Structure with its Inuring Priorities

Priority	Location	Policy	Account	Portfolio
Level 1	Surplus Share	Working Excess	Quota Share	Cat Reinsurance
Level 2	Facultative	Surplus Share	Working Excess	Corporate Cat

All the different forms of reinsurance contracts result in several combinations that typically require complex algorithms in their implementation. These algorithms relate three essential components in the model: The analysis resolution, the allocation of the insurance and reinsurance contracts, and the attribution of the loss between the participants (i.e., the insured, the insurer and the reinsurer). Looking at it from an insurer standpoint, as the model goes through the calculations, four different quantities of loss are calculated for each postulated event:

- **Ground-Up loss**: This is the total economic loss to the portfolio for the coverage under consideration
- **Gross Loss**: This is the loss after deductible and limits. The reminder represents the loss to the insured
- **Pre-Cat Net Loss**: This is the loss after all reinsurance contracts but prior to the application of catastrophe treaties. At this stage the loss to each reinsurance contract is also captured
- **Net Loss**: This is the net loss to the insurer after all reinsurance contracts and treaties are applied. At this stage the loss to the treaty reinsurer is also calculated.

A catastrophe model can output anyone, a combination, or all of these loss components depending on the need of the user. The nature of the resolution in all the dimensions discussed above (i.e., data, analysis and reinsurance modeling) impacts the uncertainty in the model. This, in turn impacts the accuracy of the model. As explained below (Section 3.4), the modeling of uncertainty is a complex process, but it is core in determining each of the loss components listed above. Catastrophe models ability to track insurance and reinsurance contracts and rationally calculate the losses to each participant, represents a significant progress in insurance practice and has introduced significant transparency and efficiency in the industry.

The issue of the data resolution is germane to catastrophe modeling. Data resolution is not any better in a global insurer/reinsurer portfolio than it is in domestic insurers from emerging economies, and there is more disparity in data between insurers and reinsurers than between insurers of different countries. Modeling companies have significant experience in dealing with varying data resolutions, and typically go through a process of data auditing and data quality control before performing an analysis. Inference rules and inventory databases used by models to improve the poor resolution of the data for a reinsurer would work as well for processing the portfolio of a domestic insurer in countries such as Turkey, India or Mexico, for example. While poor data quality increases the uncertainty in the results, catastrophe modeling lay a transparent and robust logic by which all the participants can understand their risk profile. This logic improves the quality control in data collection and processing for insurers in emerging economies, resulting in increased efficiency.

2.2 Hazard Module

The Hazard Module uses the seismic hazard information such as geology, seismo-tectonics, seismicity, soil conditions, and earthquake energy attenuation characteristics to calculate the potential for ground motion and ground failure and their related probabilities at a site. The methodology for calculating ground motion hazard is well established in the literature (Hanson and Perkins, 1995; Frankel et al., 1996). First, the seismic sources for the region are defined in terms of their geographical areas, and their related recurrence relationships, and maximum magnitudes. Sources can be area sources, line sources or point sources depending on the amount of information known on the source. Second, one or more ground motion attenuation formula is selected to define the source-to-site energy decay relationship as a function of magnitude and distance. The amplitude of the ground motion intensity parameter at each site (typically peak ground acceleration - pga) is calculated for the postulated earthquake magnitude using the attenuation relationship.

An attenuation formula applies for a specific soil condition (e.g., rock, or stiff soil) (Ambraseys and Bommer, 1995) and yields a ground motion intensity parameter relative to that particular soil condition. Factors are introduced to correct the value of the ground motion parameter to represent the soil conditions at the site of interest (Borcherdt, 1994). To perform such correction, models rely on internal and soil databases that are developed from geologic maps, and that classify the geological structures into idealized surficial soil classes. Empirical relationships prescribe the correction needed for each type of soil.

Models have additional databases that rate the landslide susceptibility and the liquefaction susceptibility of a site. (Liquefaction is a process of the transformation of a granular material from a solid state into a liquefied state as a consequence of increased pore water pressures caused by earthquake shaking (Youd and Perkins, 1978).) These databases are developed either from maps when they are available or from internal algorithms that combines the soil conditions with slope or water table considerations. These databases are used to calculate potential damage due to landslide and liquefaction.

The hazard module also contains a methodology for forecasting the occurrence of earthquake events. The most common and simplest earthquake occurrence model is the Homogeneous Poisson Model, which assumes that earthquake events are temporally independent and spatially independent (Cornell, 1968; Cornell and Winterstein, 1988). As a result, the probability of occurrence of an event is not dependent on the time of the last event. Similarly, the probability of occurrence of two events at the same time and at the same place is negligible. To represent "simulated" event, a simple process of digitizing the sources into segments and generating events by small magnitude increments on each segment is typically used. The common Gutemberg-Richter formula of the form $logN = a + bN$ (Gutemberg and Richter, 1944) provides the recurrence relationship of the events. In this relationship, N is the number of events greater or equal to M per unit time and unit area, and a and b are regression parameters based on the seismicity distribution of the source. Slip rate data, when available are used to determine the rate of large events (Wesknousky, 1986; Barka, 1988). A standard magnitude-rupture relationship is used to determine the length of the rupture. More sophisticated model that introduce time-dependency are also used when information is available on last event and typical recurrence time of "characteristic" earthquake (Schwartz and Coppersmith, 1984; California Working Group, 1988, 1990, 1995).

To develop the probability distribution of the ground motion amplitude for a site, two approaches are typically used. The traditional uniform hazard approach (McGuire, 1993) which integrates over magnitude and distance the product of the probability distribution for magnitude and the probability distribution for distance. This double integration results in a cumulative hazard curve representing the probability of exceedence of a given ground motion for any given time t. Another approach consists of running each event separately, ordering the resulting amplitude by descending order, and developing the cumulative distribution function by cumulating the probabilities. The distinct advantage of the latter is that it keeps track of the event making the integration of the losses from site to portfolio straightforward. The expected loss to the portfolio is simply the sum of the expected loss to each location for each simulated event. Its drawback is that it is portfolio dependent and thus, needs to be run real-time impacting model performance. The former method on the other hand, is portfolio independent and its results can be pre-programmed and stored in databases saving computing time. However, because the association of the event is lost, it requires some form of a correlation matrix in order to aggregate losses from each site to a portfolio. In risk transfer applications, this feature is critical because one often needs to aggregate portfolios across territories in order to represent the loss to an insurer or reinsurer book of business. For this reason, the more transparent event-based approach is generally preferred.

There is a wealth of information on active tectonics, seismicity and geological information worldwide that is used for the development of hazard models. Most, if not all the active areas of the World have been extensively studied, and hazard maps covering the active areas of the globe have been published (GSHAP, 1999). Many studies have covered the Eastern Mediterranean and the Balkan Region (Erdik, et al., 1985; Gulkan and Yuceman, 1991; Erdik et al., 1999; Papaioannou and Papazakos, 2000). These studies offer excellent references for the development of hazard models for the purpose of assessing catastrophe losses to emerging economies.

2.3 Vulnerability Module

The vulnerability module consists of a series of databases and algorithms for calculating damage quantities to the properties at risk. The databases include:

- Construction classification databases which classifies the built environment into generic construction classes
- Occupancy classification representing typical functional occupations
- Damage-probability databases that provide the probability of being in a given damage state for different levels of ground motion severity
- Inventory databases that provide a distribution of the inventory by construction class

The nature of the data stored in these databases typically varies by region and by country. Traditionally, the insurance industry has captured information on building construction according to the so-called ISO Classification, which is based on fire attributes. Insurers are updating their construction data classification to use more engineering-based classifications such as the ATC-13 classification for which the damage-intensity relationships are known (ATC, 1985). This classification provides damage probability matrices for more than 70 construction classes with about 30 of these classes pertaining to buildings. In addition, for buildings, it differentiates between low-rise (3 stories or less), medium-rise (4 to 7 stories) and high-rise (more than 7 stories). While, this classification is not universal, many of the ATC-13 construction classes have been adopted to other regions and countries.

Table 2: Sample Probability Matrix from ATC-13

Central Damage Factor	Probability						
	Modified Mercalli Intensity (MMI)						
	VI	**VII**	**VIII**	**IX**	**X**	**XI**	**XII**
0	**2.7**	*	*	*	*	*	*
0.5	**65.8**	**10**	**1**	*	*	*	*
5	**31.5**	**89.7**	**88**	**34.5**	**3.5**	*	*
20	*	**0.3**	**11**	**63.4**	**76.2**	**17.5**	**3.7**
45	*	*	*	**2.1**	**20.3**	**74.5**	**68.3**
80	*	*	*	*	*	**8**	**28**
100	*	*	*	*	*	*	*
MDR	**1.9**	**4.6**	**6.6**	**15.4**	**24.6**	**43.4**	**53.9**

The damage probability matrices of the ATC-13 are intensity-based. They provide the probability of being in one of seven damage states (spanning from no damage to total damage) as a function of Modified Mercalli Intensity (or MMI). An example of damage probability matrix is shown in Table 2. These damage probability matrices were developed based on expert opinion using a Delphi technique. Each damage state is centered on a median value. By summing the product of the probability and the medians across all damage states, one obtains the Mean Damage Ratio (or MDR) at the intensity of interest. By definition, the MDR is the cost of repair over the cost of replacement. In addition to the damage probability matrices, ATC-13 provides statistics associated with these matrices for which an estimation of the uncertainty around the MDR can be estimated. The curve that shows the variation of MDR versus MMI is referred to as Vulnerability Function. Examples of vulnerability functions are shown in Figure 2.

Figure 2: Example of Vulnerability Functions

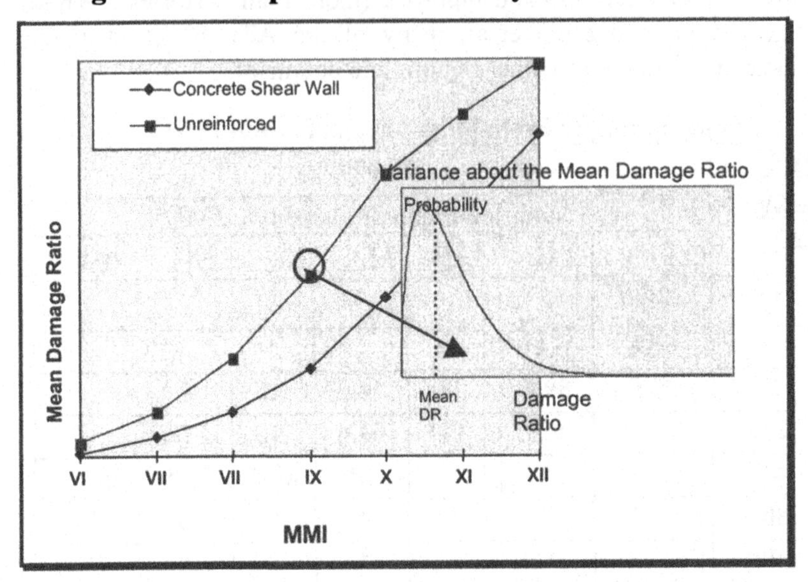

Catastrophe models use the vulnerability functions to compute the damage given a ground motion severity and the class of structure. Damage estimates are generated using separate vulnerability functions for the three main types of exposure recognized by the insurance community: Building, Contents, and Time Elements/Business Interruption. These vulnerability functions account for the varying manner in which loss occurs to the different types of exposure. For example, content losses may be more significant than building losses in a high rise structure due to building deformations. Similarly, business interruption may constitute the most important losses to a distribution facility that relies on transportation and lifelines network for its business.

Models get frequently updated to reflect more current knowledge on vulnerability of buildings and to incorporate actual data from recent earthquakes. Current state-of-the-art is moving from intensity-based vulnerabilities to spectral-based vulnerabilities in the calculation of damage (Kircher, et al., 1997). These newer approaches are gradually implemented in commercial catastrophe models. However, they are more difficult to implement because they involve more complex calculations than the intensity-based approaches. They also require more data about the construction type and the soil characteristics and the site. Finally, they are more difficult to calibrate because damage surveys have traditionally been correlated with intensity.

The analytical part of the Vulnerability Module consists of algorithms that calculate the expected level of property damage using the ground motion and probability of ground failure parameters provided by the Hazard Module for each location. Property damage is quantified as a mean damage ratio (MDR), representing the ratio of the cost to repair earthquake damage over the replacement cost. An estimate of the uncertainty around the mean damage ratio, expressed by the coefficient of variation (CV) is also calculated. Using the MDR and the CV a two-parameter distribution of the damage can be developed. Typical distributions used to represent damage include the Lognormal distribution and the Beta distribution. This distribution is passed on to the Financial Module to transform damage into **ground-up** loss and then apply the financial structures imposed by insurance and reinsurance policies to calculate gross loss and net loss.

The ATC-13 vulnerability data was developed to reflect California construction practice and experience. The concept of the ATC-13 can be adapted to other regions by enlisting the participation of local experts. Many emerging economies such as Turkey, Mexico, Philippines and others, have very competent experts who are familiar with these techniques and who can provide specific information on the building classification and vulnerability. In addition, the seismic experience in these countries is often much longer and much richer in data and investigations than the United States. For example, Turkey has excellent damage data and post-earthquake investigation information from recent Turkish earthquakes. Hence, the development of the vulnerability functions for models geared at emerging economies should not create any particular difficulty.

2.4 Financial Module

This module translates damage distributions obtained in the vulnerability model into loss distributions taking into consideration the uncertainties associated with hazard, vulnerability and exposure. This module also calculates the loss to each participant (i.e., insured, insurer, and reinsurer) according to the insurance and reinsurance policies that are in place and then aggregates the losses geographically and within each coverage. The loss distribution accounts for the possibility of the loss reaching different levels of an insurance and reinsurance policy structure. For example, on an expected value basis, a mean loss lower than the deductible level indicates a zero loss to an insurer. However, because of the uncertainties introduced in the calculations, the actual loss may be higher than the mean and may exceed the deductible registering a loss to the insurer. Similarly, the mean gross loss maybe lower than the attachment point of reinsurance contract indicting a zero loss to a reinsurer. There is always a finite probability that the actual loss is higher than the mean, thus causing a loss to the reinsurer, which on an expected value calculation would have been zero. By accounting for the uncertainty, the model integrates over the loss distribution at each stage of the calculations to find the losses to each party that is participating in a particular risk transfer contract. No matter what the mean value is, the distribution will always show that the participants will have a share in the loss according to the nature of the probability density function of the loss. Hence, in the financial module, loss calculations are a function of the value of the mean loss, the amount of uncertainty and the nature of the distribution.

Typically a Beta distribution or Lognormal distribution is suitable for representing the ground-up loss. The Beta distribution, in particular has the convenience of being bounded between 0 and 1 and is completely defined by the mean and CV. For convenience, insurance and reinsurance policies are represented as a block diagram that is coupled with the loss distribution as shown in Figure 3. This highlights the relationship between the loss distribution and the allocation of loss among the risk holders involved in the policies.

Figure 3: Loss Distribution and Reinsurance Structure

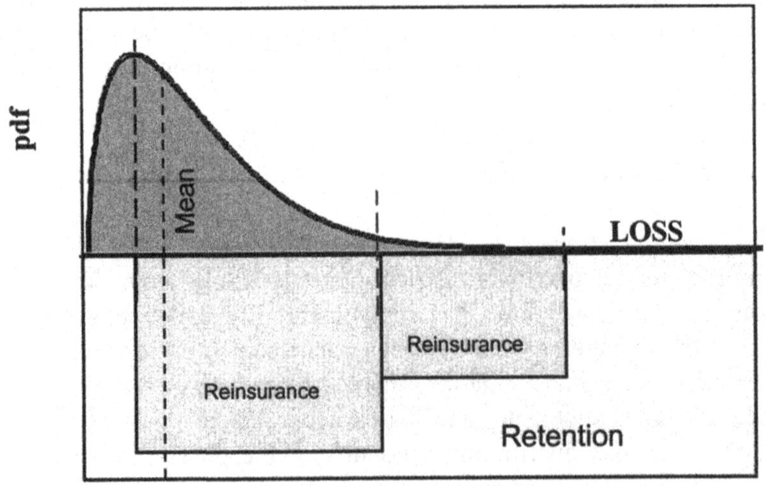

The first step in the calculation of the uncertainty[4] is to develop an aggregated CV that captures the different uncertainties that encompass the loss of a given property. Given an event with known magnitude, location and rupture has happened, the main sources of uncertainty are:

[4] The word "uncertainty" is used here to include both randomness and uncertainty itself.

- uncertainty in the quality of the underlying exposure data
- uncertainty in the attenuation formula
- uncertainty in the soil conditions
- uncertainty in the vulnerability function

Uncertainty includes both model uncertainty and parameter uncertainty. Generally, for any given event and for a given property, these sources of uncertainty are considered to be independent. However, as the calculations proceed, several assumptions and additional techniques need to be introduced in order to keep track of uncertainty in an appropriate manner. For example, the assumption of a Beta distribution for the ground-up loss may not apply to the net loss of a particular layer in loss allocation structure. The distribution of any particular layer of reinsurance will be truncated between the trigger value and the exhaust value. Similarly, the aggregation of loss from a single location to multiple locations (i.e., account or portfolio) changes the nature of the loss distribution (i.e., the sum of Beta distributions is not necessarily a Beta distribution itself.) and requires additional corrections. As the calculations proceed, the formulation of the parameters of the distribution may become difficult, if not impossible to resolve analytically, and numerical techniques such as simulation are introduced to approximate uncertainty parameters.

Another important parameter in the determination of uncertainty is the amount of correlation between locations. Correlation is related to factors such as the type construction, or the geographical distribution of the locations. As one aggregates the ground up losses from locations into a portfolio, correlation between locations impacts the uncertainty. If one assumes that the locations are completely independent, then the aggregate variance is equal to the sum of the variances for each location. On the other hand, if one assumes that the locations are completely correlated, then the standard deviation is equal to the sum of the standard deviations at each location. Hence, correlation increases uncertainty. For example, if the properties in the portfolio have all the same construction and are concentrated in the same area, the uncertainty is greater because a small error in the damage calculation or the ground motion calculation (or both) would result in large errors in the portfolio loss. Assumptions on correlation are generally based on empirical data and sensitivity analyses.

The end results from the Financial Module calculations consists of a mean and a CV for each of the portfolio **ground-up** loss, **gross** loss, **pre-cat** net loss and **net** loss for each simulated event. An additional layer of uncertainty is then considered at this point. It relates to the uncertainty in the rate of occurrence of the event and in the model itself. This type of uncertainty is generally referred to as "primary" uncertainty, as opposed to the uncertainty in loss given an event, which is referred to as "secondary" uncertainty. Clearly, the rate of occurrence of earthquake events (especially large infrequent events) is highly uncertainty. Similarly, the model one uses to calculate the rate or the loss is not perfect. These uncertainties get incorporated after the loss given an event is calculated. For example, uncertainty in the rate intervenes when calculating the portfolio loss corresponding to a given probability, or the probability of reaching a given portfolio loss. Model uncertainty is either neglected, or dealt with by using more than one model and averaging the results following a logic tree.

3. MODEL OUTPUT AND APPLICATIONS

Typically, a model provides several analysis capabilities that can generate loss estimates specifically tailored to different types of risk management decision making. Scenario analyses can be used to generate "what-if" losses for repeats of historical earthquakes or for a postulated future event. The deterministic analysis is of great benefit to understand the human and economic consequences of particular earthquakes on a particular region or portfolio (RMS, 1995 a, b, and c). Probabilistic analyses indicate potential portfolio losses for a range of return. Two types of outputs are generated from a probabilistic analysis: Average Annual Loss (AAL) and Loss Exceeding Probability (LEP).

AAL represents the expected long-term cost of the risk. It is defined as the sum over all events of the product of the loss and the rate of occurrence of the loss. Therefore, it incorporates both the frequency and severity of loss. When normalized to a unit of exposure, the AAL represents the "pure premium" or "technical premium". In terms of gross loss, the pure premium represents the rate that an insurer needs to charge in order to equally balance the long term risk of the property that is being insured. This applies to any layer of loss under consideration. Hence, AAL output is used to set up pure premiums that serve as a basis for establishing insurance rates or tariffs.

Figure 4: Illustration on the Use of Average Annual Loss to Examine Portfolio Efficiency

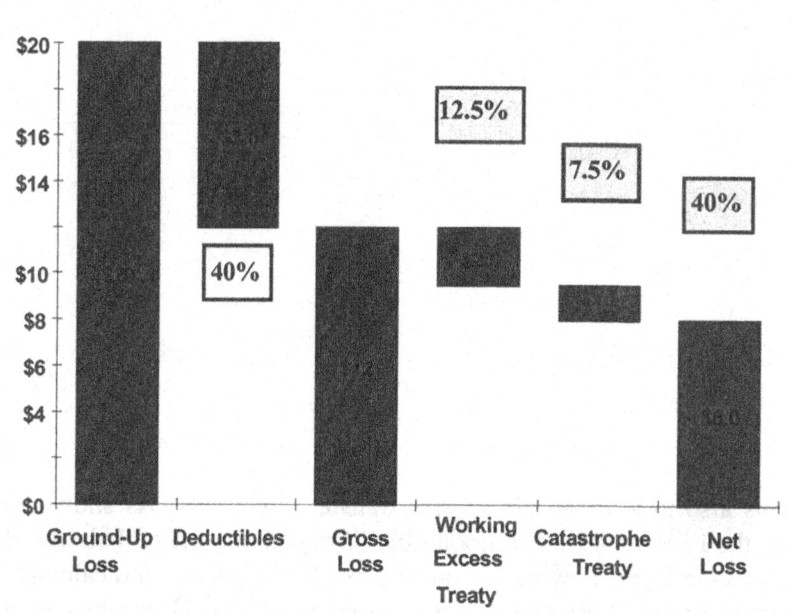

From portfolio analysis purposes, an analysis of the AAL can be used to determine the relative risk of various components in a portfolio, including policies, geographies and construction classes. An AAL analysis is also useful to evaluate the efficiency of an insurance or reinsurance structure. A hypothetical case is shown in Figure 4. Starting from a ground up AAL of $20 million, one can see that this particular portfolio gets a 40% dissipation from the deductible and 12.5% and 7.5% dissipation from the two reinsurance contracts. These measures can be contrasted with the price paid for each contract. The indication in this example is that the deductible provides great protection to the portfolio whereas the reinsurance contracts provide rather weak protection.

The LEP curves present the annual probability that a given exposure will undergo a specific level of loss. The advantage of generating such a loss quantification curve is far reaching. First, the LEP allows a risk taker (insurer, reinsurer, etc) to establish measures for tracking risk in a consistent manner. By looking at losses at specific return periods, potential exposure to catastrophes can be evaluated by comparing it to surplus or premium collected. Similarly, concentrations can be controlled in this manner. An example of such analysis is shown in Table 2. Companies can use this analysis to evaluate the impact of change in underwriting practices, or to design targeted portfolio strategies to mitigate negative trends.

Table 3: Example of Use of Results from a Catastrophe Model to Study Portfolio Positions with Respect to Key Measures such as Limit, Premium and Surplus.

RP Years	One Year Prob. (%)	Ground-Up Loss (x1,000)	Gross Loss (x1,000)	Net Loss (x1,000)	Net Loss Percent of Limit	Net Loss Percent of Premium	Net Loss Percent of Surplus
20	5.00	$260	$110	$73	0.91%	73%	4.9%
50	2.00	$520	$250	$190	2.38%	190%	12.7%
100	1.00	$1,000	$420	$330	4.13%	330%	22.0%
250	0.40	$2,200	$1,300	$1,000	12.50%	1,000%	66.7%
500	0.20	$2,800	$1,800	$1,400	7.50%	1,400%	3.3%
750	0.13	$3,300	$2,200	$1,750	21.88%	1,750%	116.7%

LEP is also used to decide on risk transfer structures. As shown in Figure 5, the LEP provides the probability of trigger and the probability of exhaust of each layer showing various layers of reinsurance, and calculate the Average Annual Loss (AAL) for each layer. The AAL establishes a basis for the rate (pure premium) at which the risk associated with that layer can be transferred. Companies can compare the relative effectiveness of the different contracts in mitigating potential catastrophe losses. For example, they can compare pure premiums and associated standard deviations to rate on line in making reinsurance purchase or restructuring decisions.

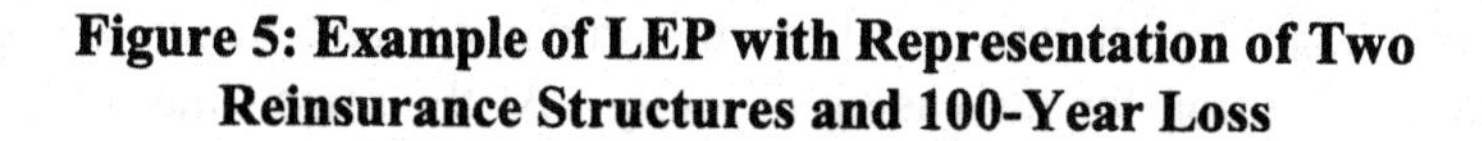

Figure 5: Example of LEP with Representation of Two Reinsurance Structures and 100-Year Loss

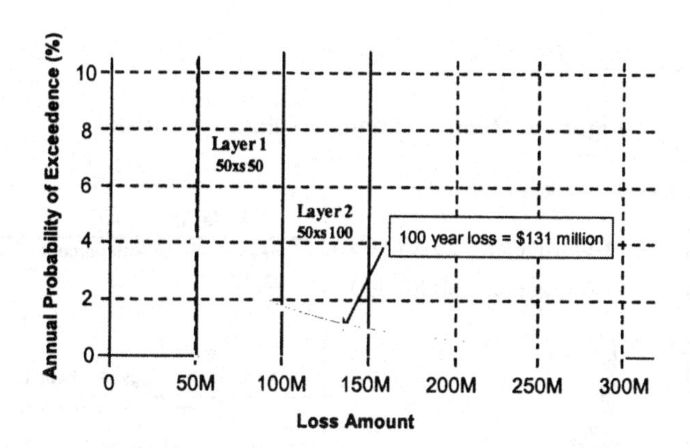

Another application of LEP is for pricing of risk. In this case, volatility needs to be introduced. Calculating the AAL at the expected value and at some confidence interval (say 90th percentile) provides an indication of the volatility of the risk. This volatility is due both to loss uncertainty and rate uncertainty (i.e., secondary uncertainty and primary uncertainty). Volatility increases the price of a risk transfer transaction because investors are less willing to assume highly volatile risks. For high severity low frequency risks such as earthquakes, the volatility naturally increases towards the end of the distribution mainly driven by rate uncertainty associated with large events. An example is shown in Figure 6 where two contracts are considered, one with a relatively high probability of trigger of 5% (1 in 20 years) and another one with the much lower probability of trigger of 0.5% (1 in 200 years). This example shows that the volatility of the AAL and the volatility of trigger probability of the second layer are about twice as large than the first layer. The quantification of uncertainty in AAL and the probability of loss is an important parameter in the pricing of risk. For example, catastrophe bond transactions rely greatly on this type of sensitivity analysis to establish a price and rating of a bond.

The ability of models to calculate basic risk parameters and to represent the volatility in these parameters has helped expend the use of catastrophe modeling within the financial market. Investment banks are designing catastrophe investment products for the capital markets (such as catastrophe bonds). The capital market has a much larger amount of capital than traditional reinsurance companies, and hence could be an attractive option for increasing capacity. At the same time, investors are very interested by these types of investments because they are not correlated to traditional financial investments (Canabarro, et al., 1999).

Figure 6: Quantification of Volatility of Risk

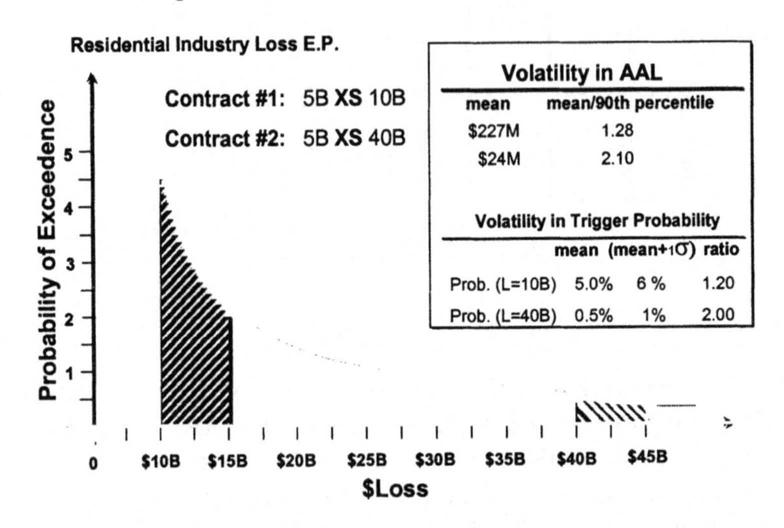

4. APPLICATION TO EMERGING ECONOMIES

Emerging economies such as Turkey, Mexico, India and others, which repeatedly get hit by large catastrophes, need to bridge the gap in technology and know-how in the field of disaster and risk management. Currently, the tremendous burden of catastrophes falls overwhelmingly on the responsibility of the government and private individuals. After each catastrophe, governments of emerging economies divert valuable development and social services resources into the rebuilding process. The cost of catastrophes keeps increasing due to intensive urbanization, aging infrastructure and lack of preparedness. Today global economy makes it even harder for emerging economies to pursue their quest for development after a major earthquake hits them. It typically takes years for a country to recover from a major catastrophe.

International funding organizations such as the World Bank are pushing for reforms in developing countries to shift the burden of catastrophes from the public sector into the private sector. One such reform is implementation of compulsory catastrophe insurance. In September 2000, a decree-law went into effect in Turkey making catastrophe insurance for private dwellings mandatory. Countries such as Greece and others are also considering establishing some form of mandatory insurance. The effects of such laws on both government and the insurance industry are far reaching. The government needs to make sure that the process of shifting the costs of catastrophes to the insurance industry actually works. Insurers will only be willing to increase their exposure if they can charge adequately for the risks they are taking, and if they can find the reinsurance capacity to protect them from future losses. Hence, without adequate rates and sufficient capital, the actual implementation may not yield the expected results, and the transfer of liability may not happen. The nature of this trade-off is better understood and remains transparent to all parties when catastrophe models are used to quantify the risk parameters. Transparency is key to building confidence among all stakeholders. A compulsory insurance scheme that works like a tax and does not account for the realities of the underlying risk parameters may just simply be an additional burden on all parties.

Implementation of mandatory laws increases insurance penetration and results in larger concentration of risk. Even with adequate rates and the availability of reinsurance capacity, the management of these concentrations must be done rationally in order to protect policyholders. The compulsory insurance program of Turkey, referred to as TCIP (Turkish Catastrophe Insurance Pool) is centrally managed by a Board, which contracts with a Turkish Reinsurance company (Milli-Re) to manage the portfolio. The capacity for reinsurance is in great part, provided through the global reinsurance market. If and when the TCIP portfolio grows to its full potential, it would turn Turkey into an attractive market for the global reinsurance market. At the same time, the underlying risk to the portfolio is significant because many of Turkey's large cities (e.g., Istanbul, Izmir, Adana, etc.) are located in highly seismic areas. Without the benefit of analytical tools embedded in catastrophe models, it would be difficult to manage such a complex portfolio. Catastrophe models would specify adequate rates, estimate realistic probable maximum losses, and indicate optimum risk transfer structures.

The adoption of catastrophe modeling technology in emerging economies must overcome the misconception that catastrophe modeling "*is only for developed markets*". First, there are very few scientific and technological hurdles that need to be overcome to build an effective catastrophe model. Most often, the knowledge in seismology, vulnerability and other sciences within institutions of emerging economies is competitive with the one in developed countries. The local experts could supply the scientific information needed to build a model. If there is a lack of knowledge, it is most probably in putting the technology together. The support of modeling companies from abroad can overcome that hurdle. Second, the quality of the exposure data is not necessarily worse than in developed countries, and can easily be dealt with by models. Third, the cost of development of technology is relatively low, especially if local experts are used to help with the development. The most significant challenge is the investment in the organizational skills and infrastructure to support and effectively use the technology. Organizations will gain little benefit from modeling technology if they do not develop the internal infrastructure to support the application of the models, and carry the results to the decision-makers within their organization.

The use of catastrophe modeling is having a profound effect on the way risk is quantified, managed and traded on the global financial market. Underwriters have been making calculated decisions regarding catastrophe risks for years, based on their experience and whatever tools are available to them. This remains the current practice in emerging economies. However these decisions can be made more effectively with improved analysis, leading to lower uncertainty, more efficient transactions and greater dynamics in the market. In today's global market, financial institutions in emerging economies must acquire the technological tools and know how to compete; otherwise they will loose market share and risk insolvency.

Similarly, government organizations in charge of disaster and risk management must acquire the technological tools by which they can measure the impact of earthquakes on their populations, critical services and infrastructure. These tools will help institutionalize the application of disaster management in the day-to-day operations of urban planning, construction monitoring and disaster preparedness. Government organizations in emerging economies have made progress in improving their disaster response capabilities. While these actions are critical to respond to the needs of the victims, they have little impact on reducing the human and economic losses of disasters. Pro-active actions of preparedness and mitigation are the only sure ways to reduce future losses. The use of technology will improve the planning capabilities by providing realistic loss scenarios, indicating vulnerabilities and weak links, and estimating resources. Moreover, the discipline of a model will force both private and public institutions involved in risk management to seek and improve their knowledge in the field, to improve their data collection processes, and to impose rigor and transparency in their risk management practices. This in turn will reduce the volatility of their losses, decrease the costs of operations and increase the efficiency and credibility of their organizations.

5. CONCLUSION

Computer modeling has considerably enhanced the capabilities of insurance, reinsurance and other financial companies to quantify manage and transfer seismic and other catastrophe risks. Despite the uncertainties associated with models, they provide substantial value in the form of unprecedented insight at company exposure, loss drivers in a portfolio, and efficiency in risk transfer structures. Insurers and regulators in emerging economies can use catastrophe modeling for determining rates, deciding on risk transfer options, establishing underwriting guidelines, deciding on capital allocations, measuring financial performance, establishing risk management policies, and determining marketing strategies. This would result in increased transparency and efficiency in their organizations.

Similarly, catastrophe modeling is a powerful tool for emergency managers and city planners to understand impacts of earthquakes and other disasters, and to develop disaster response, and preparedness plans. Urbanization is growing at an alarming rate in emerging economies. Cities such as Istanbul, Manila and Mexico City exceed more than 10 million inhabitants and keep growing. The management of such cities is increasingly challenging diverting attention of public officials towards day-to-day issues. However, these cities are at high risk from earthquakes and other disasters, and could suffer losses of incredible magnitude. The most effective way to reduce the risk of disasters to large cities is to institutionalize disaster management within their organizations and operations. Catastrophe models represent powerful tools for mapping the impacts of disasters and for understanding the consequences on population, essential facilities and services and infrastructure of a city. This information can be turned into planning parameters, and into decision-making tools related to urban planning, building code standards, and preparedness programs for populations and institutions.

Fortunately, there are some encouraging recent developments in this area. Various megacities from emerging economies are acquiring and developing technologies for disaster management. The Municipality of Metropolitan Istanbul is completing the preparation of an earthquake management master plan centered on earthquake loss assessment technology. Other cities such as Manila, Mexico and Santa Fe de Bogota are also experiencing with similar technology. In addition, many cities of emerging economies are rapidly increasing their knowledge by working with international organizations and partnering with their own academic institutions to advise them on best practice and to help them develop the required resources (Bendimerad, 2000). Putting catastrophe models in the hand of the city planners and the risk underwriter would provide the institutional knowledge that allows private and public institutions in emerging economies to build the internal capacity for effectively quantifying, managing and reducing losses from disasters.

REFERENCES

Ambraseys, N.N. and J.J. Bommer. 1995. "Attenuation Relations for Use in Europe: An Overview." In *European Seismic Design Practice*, edited by Elnaishai, Rotterdam: Balkema Publications.

Applied Technology Council. 1985. *Earthquake Damage Data for California (ATC-13).* Redwood City, CA.

Barka, A.A. 1988. "Strike-Slip Fault Geometry and its Influence on Earthquake Activity." *Tectonics*, Vol. 7, No. 3: 663-684.

Bendimerad, F. 2000. "Megacities-Megarisk." In *Managing Risk. A Special Report on Disaster Risk Management*", Provention Consortium, World Bank, Washington D.C.

Borcherdt, R.D. 1994. "New Developments in Estimating Site Effects on Ground Motion." *ATC-35-1*, Applied Technology Council, Redwood City, California.

Cameron, A., 2000. "Better than a Cure." *In Managing Risk. A Special Report on Disaster Risk Management.*" Provention Consortium, World Bank, Washington D.C.

Canabarro, E., M. Finkemeier, R. Anderson, and F. Bendimerad. 2000. "Analyzing Insurance-Linked Securities." *Journal of Risk Finance* Vol. 1, No. 2, N.Y. (Also in Financing Risk and Reinsurance, Sept. 1999. International Risk Management Institute, Dallas, TX.)

Carter, R.L., L. Lucas, and N. Ralph. 2000. *Reinsurance*. Reactions, London.

Cornell, C.A. and S. Winterstein. 1988. "Temporal and Magnitude Dependence in Earthquake Recurrence Models." *BSSA*, Vol. 78.

Cornell, C.A., 1968. "Engineering Seismic Risk Analysis", BSSA, Vol. 58.

Erdik, M., V. Doyuran, N. Akkas and P. Gulkan. 1985. "A probabilistic Assessment of Seismic Hazard in Turkey." *Tectonophysics*, 117: 295-344.

Erdik, M., Y.A. Biro, T. Unur, K. Sesetyan and G. Birgoren. 1999. "Assessment of Earthquake Hazard in Turkey and Neighboring Regions." *Annali de Geofisica*, Vol. 42, (No. 6, December): 1125-1138.

Federal Emergency Management Agency (FEMA). 1994. Assessment of the State-of-the-Art Earthquake Loss Estimation Methodologies - FEMA 249. Washington D.C.

Frankel, A., C. Mueller, D. Perkins, E.V. Leyendecker, N. Dickman, S. Hanson and M. Hopper. 1996. *National Seismic Hazard Maps: Documentation June 1996.* Open-File Report 96-532, USGS, Denver.

Giardini, D. (Editor). 1999. Global Seismic Hazard Assessment Program (GSHAP) 1992-1999, Summary Volume. Annali di Geofisica, Vol. 42, No. 6.

Gulkan, P. and M. Yuceman. 1991. *Practice of Earthquake Hazard Assessment in Turkey.* Report Prepared in Relation with the IASPEI-ESC Project on Worldwide Seismic Hazard Applications, Dept. of Civil Engineering, Dept. of Statistics, Middle East Technical University, Ankara, Turkey.

Gutenberg, B. and C.F. Richter. 1944. "Frequency of Earthquakes in California." *BSSA*, Vol. 34.

Hanson, S., and D. Perkins. 1995. Seismic Sources and Recurrence Rates as Adopted by USGS Staff for the Production of the 1982 and 1990 Probabilistic Ground Motion Maps for Alaska and Conterminous United States. USGS Open File Report 95-257.

Kiln, R. 1981. *Reinsurance in Practice*. Witherby and Co., Ltd, London.

Kircher, C. A., W. Holmes, O. Kustu, A. Nassar, R. Reitherman, and R. Whitman. 1997. "Building Damage and Loss." *Earthquake Spectra*, Special volume on Loss Estimation, Earthquake Engineering Research Institute, Oakland, CA.

McGuire, R. (Editor). 1993. *The Practice of Earthquake Hazard Assessment*. International Association of Seismology and Physics of the Earth and European Seismological Commission.

National Institute of Building Sciences – NIBS. 1997 and 1999. *HAZUS Technical Manuals, 3 Volumes*. Washington, D.C.

National Research Council. 1989. *Estimating Losses from Future Earthquakes*. Report of the Panel on Earthquake Loss Estimation Methodology, National Academy Press, Washington D.C.

Papaioannou, C. A. and B.C. Papazachos. 2000. "Time-Independent and Time-Dependent Seismic Hazard in Greece Based on Seismogenic Sources." *Bulletin of the Seismological Society of America*, 90, 1: 22-33.

Risk Management Solutions, Inc. (RMS). 1995a. *What if the 1906 Earthquake Strikes Again? A San Francisco Bay Area Scenario*. Topical Issues Series, RMS, Menlo Park, CA.

Risk Management Solutions, Inc. (RMS). 1995b. *What if a Major Earthquake Strikes the Los Angeles Area?* Topical Issues Series, RMS, Menlo Park, CA.

Risk Management Solutions, Inc. (RMS). 1995c. *What if the 1923 Earthquake Strikes Again? A Five Prefecture Tokyo Region Scenario*. Topical Issues Series, RMS, Menlo Park, CA.

Schwartz, D.P. and K.J. Coppersmith. 1984. "Fault Behavior and Characteristic Earthquakes." *J. of Geophys. Research*, Vol. 89.

Wesknousky, S.G. 1986. "Earthquakes, Quaternary Faults, and Seismic Hazards in California." *J. of Geophys. Research*, Vol. 91.

Working Group on California Earthquake Probabilities. 1988. *Probabilities of Large Earthquakes Occurring in California on the San Andreas Fault*. Open File Report 83-398, USGS, Menlo Park, CA.

Working Group on California Earthquake Probabilities. 1990. *Probabilities of Large Earthquakes in the San Francisco Bay Region, California*. Circular 1053, USGS, Menlo Park, CA.

Working Group on California Earthquake Probabilities. 1995. "Seismic Hazards in Southern California: Probable Earthquakes, 1994 to 2024" *BSSA*, Vol. 85, No.2.

Youd, T.L. and D. M. Perkins. 1978. "Mapping of liquefaction induced ground failure potential." *Proc. of the ASCE, Journal of the Geotechnical Engineering Division*, Vol. 104 (GT4): 433-446.

3

Challenges in Marine Seismology in the Aegean Region

Joanna E. Papoulia
Institute of Oceanography, National Center for Marine Research, Athens

Since 1996, the Institute of Oceanography of the National Center for Marine Research, and the Institute of Geophysics of the University of Hamburg, in cooperation with the Geodynamic Institute of the National Observatory of Athens, have initiated a series of projects involving seismic reflection and refraction for the investigation of the crustal type and structure of the Aegean. Furthermore, the microseismic activity of the area was monitored using combined seismological networks of land stations and ocean bottom seismographs. The main results of these projects are summarized in the present paper.

1. INTRODUCTION

The region of Aegean and surrounding areas in the eastern Mediterranean lies in the boundary zone between the Eurasian and African plates (McKenzie, 1970). This region is a zone of widespread deformation, within which complex relationships exist among extensional, compressional, and strike-slip deformations (McKenzie, 1978; Dewey and Sengor, 1979; Le Pichon and Angelier, 1979). This reveals high seismic activity, most of which is associated with the Hellenic Arc (Papazachos, 1990).

The Hellenic Arc - Trench system is a subduction zone of about 1000 km, where the African lithosphere is subducting under the Aegean microplate in a roughly SW-NE direction. This motion of the Hellenic consuming boundary with respect to Africa results from three different processes: the northward motion of the African plate, the Aegean extensional spreading, and the westward motion of the Anatolian plate (McKenzie, 1970, 1978). The active tectonic deformation in the Aegean and surrounding area has been studied by many authors using

P.R. Kleindorfer and M.R. Sertel (eds.), Mitigation and Financing of Seismic Risks, 41–52.

seismological, neotectonic, geomorphological, palaeomagnetic, as well as geodetic data, and many different models have been proposed (e.g. Papazachos and Comninakis, 1971; McKenzie, 1972; Taymaz et al., 1991; Le Pichon et al., 1995).

Figure 1: Plate motions, which affect active tectonics in the Aegean and surrounding area.

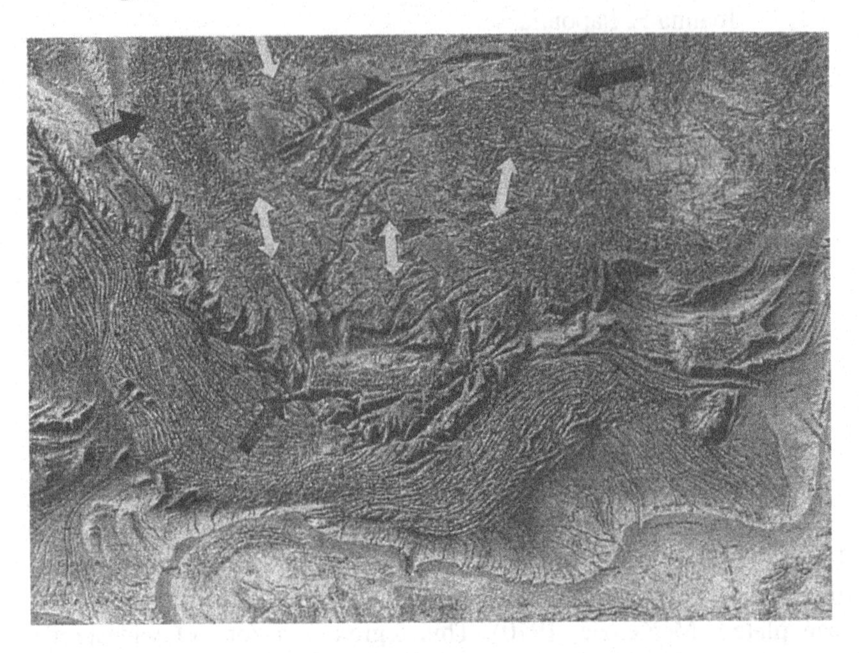

Figure 1 summarizes the geodynamic processes in the Aegean and surrounding area. Along the Apulia zone and the external Hellenic Arc reverse faulting is dominant, due to the collision of the lithospheres of Apulia and eastern Mediterranean, respectively, with the Aegean plate. The eastern Mediterranean lithosphere is subducted under the Aegean, and the associated Benioff zone shows strike-slip faulting with a reverse component. In the back-arc area normal faulting prevails along two zones, a dominant one with N-S extension, and a secondary one following the Hellenides with E-W extension. Along the North Aegean Trough and the transform zone between Apulia and eastern Mediterranean in the Cephalonia area dextral strike-slip faulting is present. The kinematic situation of the Aegean is characterized by a gradual increase of deformation velocities with respect to Europe, from 10 mm/yr in the North Aegean Trough up to 35-40 mm/yr in the southernmost Hellenic Arc. Ryan et al. (1973) estimated a subduction rate of the order of 1,5 cm/yr. McKenzie (1978) obtained a slip rate of 7 cm/yr, while Le Pichon and Angelier (1979) obtained a slip rate of 2 cm/yr for the western and 4 cm/yr for the eastern part of the Hellenic Arc, pointing out that the difference is due to a rotation pole rotated relatively close to the boundary.

Greece occupies the highest seismicity of the eastern Mediterranean, and is one of the most seismically active regions in the world. High shallow earthquakes (magnitudes up to M_S=8.3) occur mainly along the Hellenic Arc, while intermediate depth seismicity is associated with the southern part of the Aegean. The majority of these earthquakes originate from the submarine part of the Ionian Sea, the north Aegean Trough, the island of Crete, and the Dodecanese islands in the southeastern Aegean Sea (Galanopoulos, 1955; Makropoulos et al., 1989; Papazachos and Papazachou, 1989).

Seismic activity in Greece is monitored on a permanent basis by the Seismograph Network of the Geodynamic Institute of the National Observatory of Athens, consisting of 17 telemetric and 6 satellite stations. The continuing development and increase of seismograph networks within the last twenty years, has significantly improved the quantity and quality of seismic monitoring. However, the problem remains as concerned to microearthquake activity (magnitudes less than 3.0) associated with the southern part of the Aegean, where the geometry of the seismograph stations does not allow to accurately locate microearthquakes. For these cases, ocean bottom seismographs are recently used on a temporary basis to improve the geometry of the permanent network, and the recording of small magnitude events.

2. ADVANCES IN MARINE SEISMOLOGICAL RESEARCH

In the last five years, several deep seismic sounding experiments have been carried out in the Aegean and surrounding region, in an attempt to better understand its complex geodynamic evolution. A summary of the results of these investigations is presented in the following.

In Spring 1996, a wide angle reflection profiling survey was carried out in the northern Evoikos and Maliakos-northern Sporades basin, central Greece. Two seismic lines of 160 km each were observed, using 15 ocean bottom seismographs and 10 land stations (figure 2). As energy source, one sleeve gun of 60 lt volume operating at 120 bar was used. The most striking result was the detection of Moho discontinuity at a depth about 19 km below the central part of northern Evoikos (figure 3), whereas in the neighbouring mainland of Evia and central Greece the crustal thickness increases to 30 and 40 km, respectively (Makris et al., 1996; Makris et al., 2000).

Figure 2: The northern Evoikos and Maliakos – Sporades profiles. The position of the OBS's is marked by triangles.

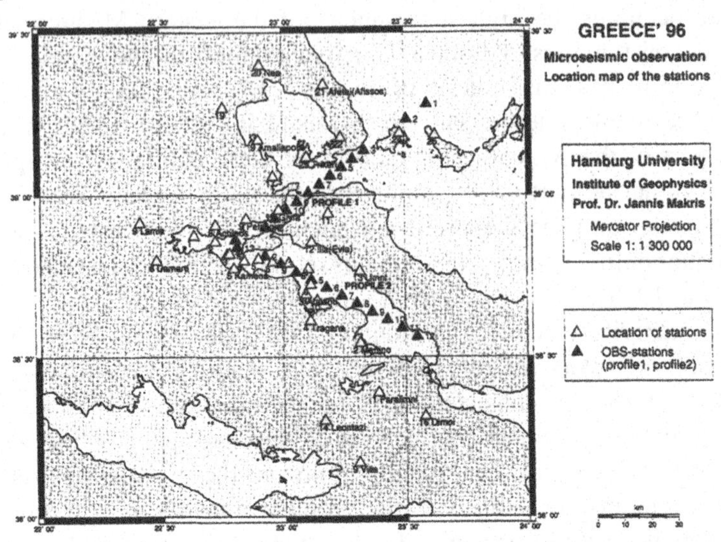

Figure 3: Velocity Model for Profile II (northern Evoikos). Moho was detected at a depth of about 19 Km in the central part of the Profile. From Markris et al. 2000.

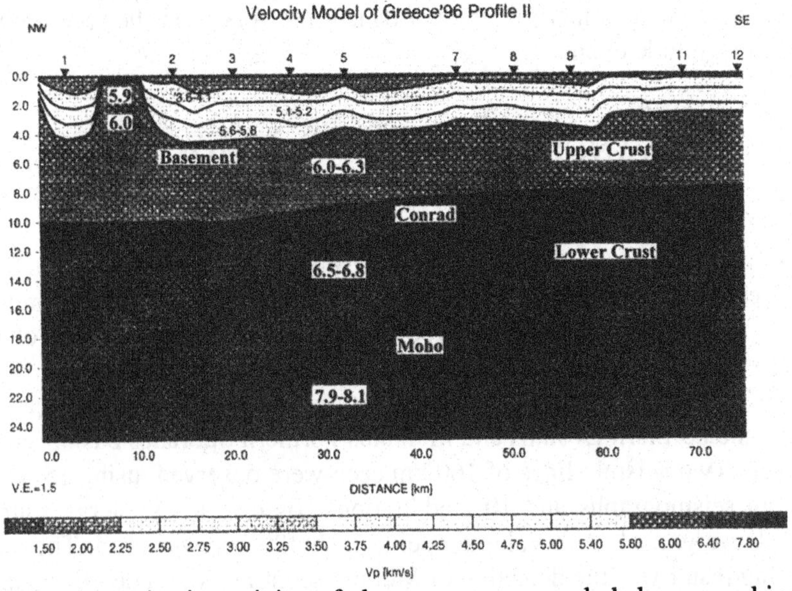

The microseismic activity of the area was recorded, by a combined network of 10 marine (OBS) and 25 land stations, operating for a two-month period. The location map of this network is presented in figure 4. During the operation of the network, a high seismic activity was observed originating from the Atalanti - Bralos area, central Greece. The network recorded approximately 1900 microearthquakes, with an average of 65 events per day, while a maximum of 333 events were reported on March 14, 1996. Most of these events were recorded by the OBSs, whereas only 450 earthquakes were detected by the land stations. The epicentral distribution of the earthquakes of the present array is presented in Figure 5.

Figure 4: Location map of stations deployed for the microseismicity monitoring (red triangles). The permanent stations of the National Seismograph network are marked in white.

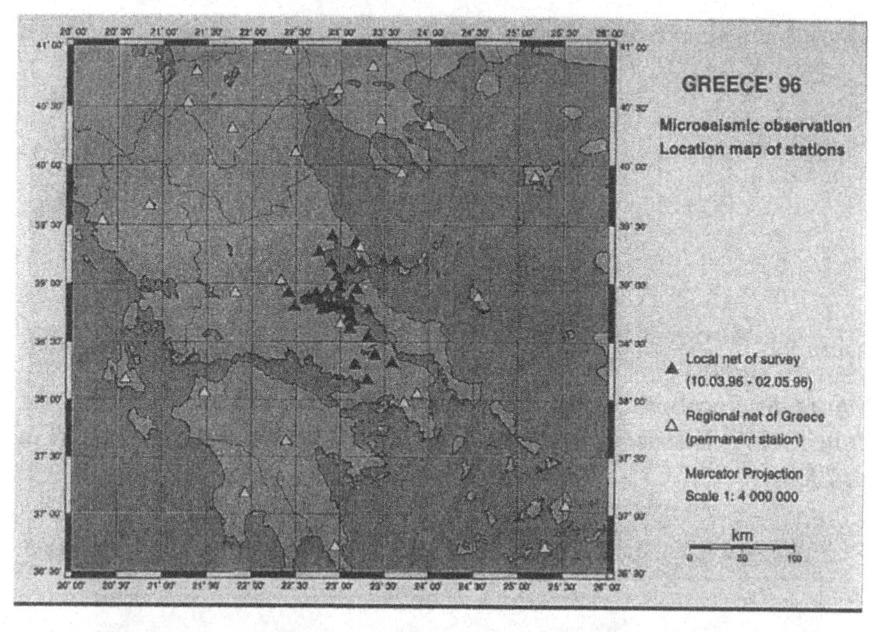

In September 1997, a project was initiated for the detailed investigation of the crustal structure in the southern part of the Hellenic Arc, around the island of Crete, and the definition of the state of collision between the African and Eurasian lithospheric plates in this area. Three onshore/offshore wide angle reflection profiling lines were shot, each of a length of about 300 km (figure 6), using 40 ocean bottom seismographs and 95 land stations. As energy source a 48 lt. airgun array, and 8 landshots of 20 kg each were used, offshore and onshore, respectively. The results of the survey were impressive, showing the complexity of the crustal structure, especially in the area south of Crete (see figure 7) (Bohnhoff et al., 2000). The oceanic crust of the subducting African plate was identified in all three profiles.

Figure 5: Location Map of Events recorded by the Local Seismograph Network of Marine and Land Stations (Period of observation: 10.03.96-02.05.96). From Makris and Papoulia (unpublished data).

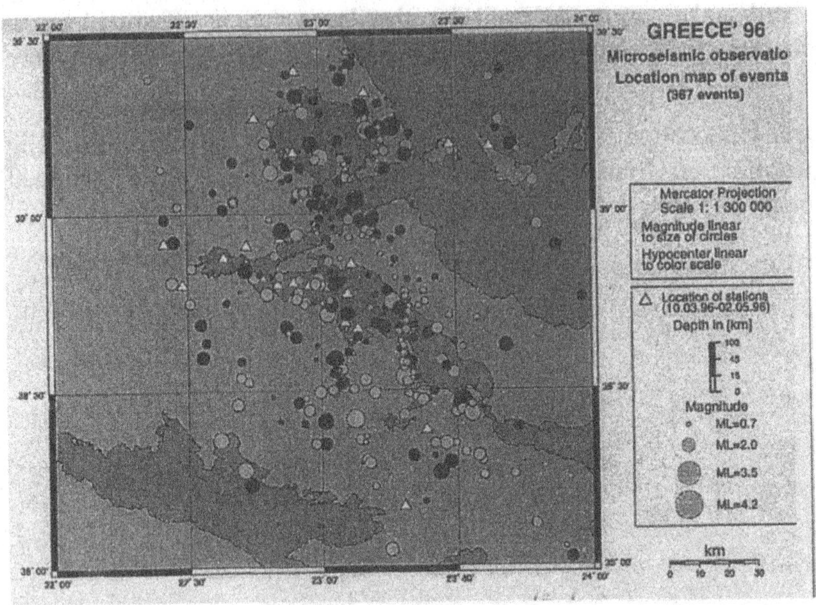

Figure 6: Location of the three seismic profiles around Crete. Crete Project 1997, Distribution of OBS- and Land-Stations. From Bohnhoff et al. 2000.

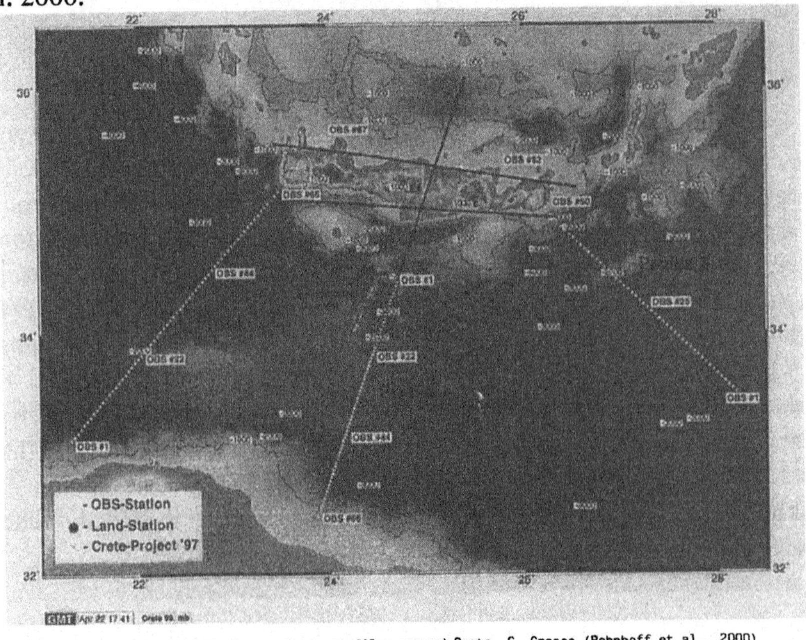

Figure 6: Location of the three seismic profiles around Crete, S. Greece (Bohnhoff et al., 2000)

Figure 7: Velocity model of profile I, NS oriented. The subduction of the oceanic crust of the African plate is clearly shown. The structure of crust appears more complex in the southern part of Crete, where intense fracturing is present. From Bohnhoff et al. 2000.

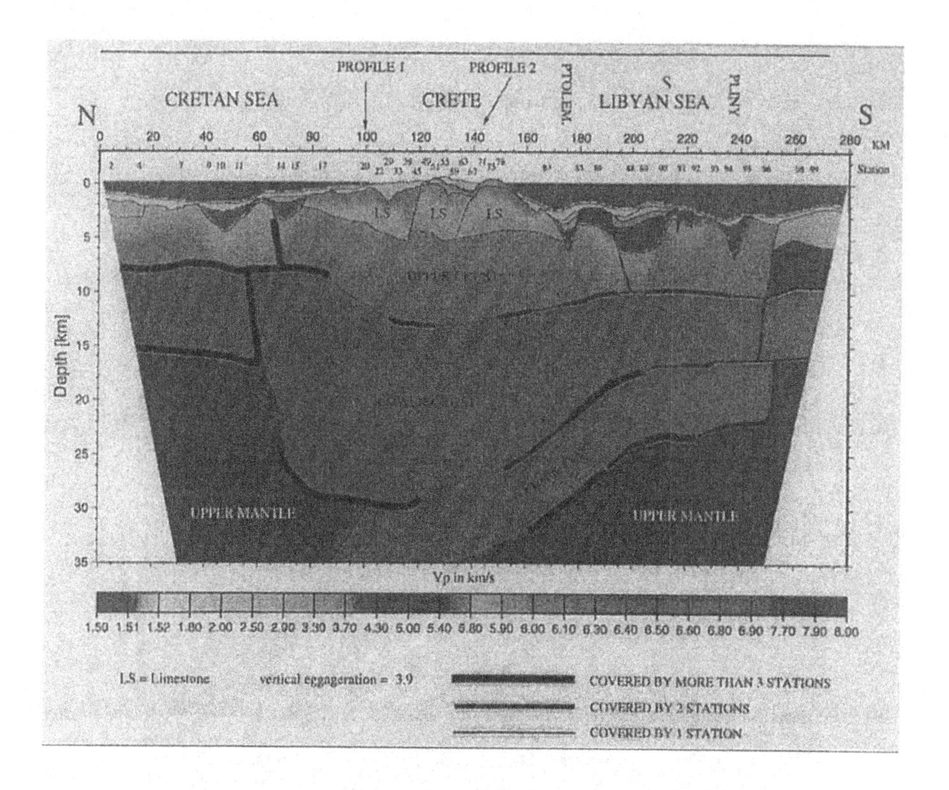

In October 1997, a seismic experiment was conducted along a 140 km profile extending from the island of Chalki to the island of Leros, crossing the volcanoes of Nisyros-Yali, in the southeastern part of the Hellenic Arc (figure 8). 20 ocean bottom seismographs and 20 land stations were deployed and recorded more than 1000 shots by a 48 lt airgun array. The seismicity of the area was recorded by deploying 22 digital stations onshore. The seismic activity was high within the caldera of the Nisyros volcano, ranging from 17 to 30 events per day. About 1200 earthquakes were located originating mainly from the upper crustal section at depths less than 10 km (figure 9).

The velocity model mapped a volcanic intrusion below Nisyros. The shallowest penetration depth of the magma was only 1.8 km below surface. The shallow depth of the intruded magma elevated the isotherms below the volcano significantly, so that the aquifers above the intrusion reached a mean temperature of 270^0 C. If the high seismic activity of that time had fractured and opened the overlaying lithological units, the overheated aquifers could have had reacted explosively, as was the case in 1873 (Makris et al., 1998), (Figure 10).

Figure 8: Orientation of the Nisyros profile. From Makris et al. 1998.

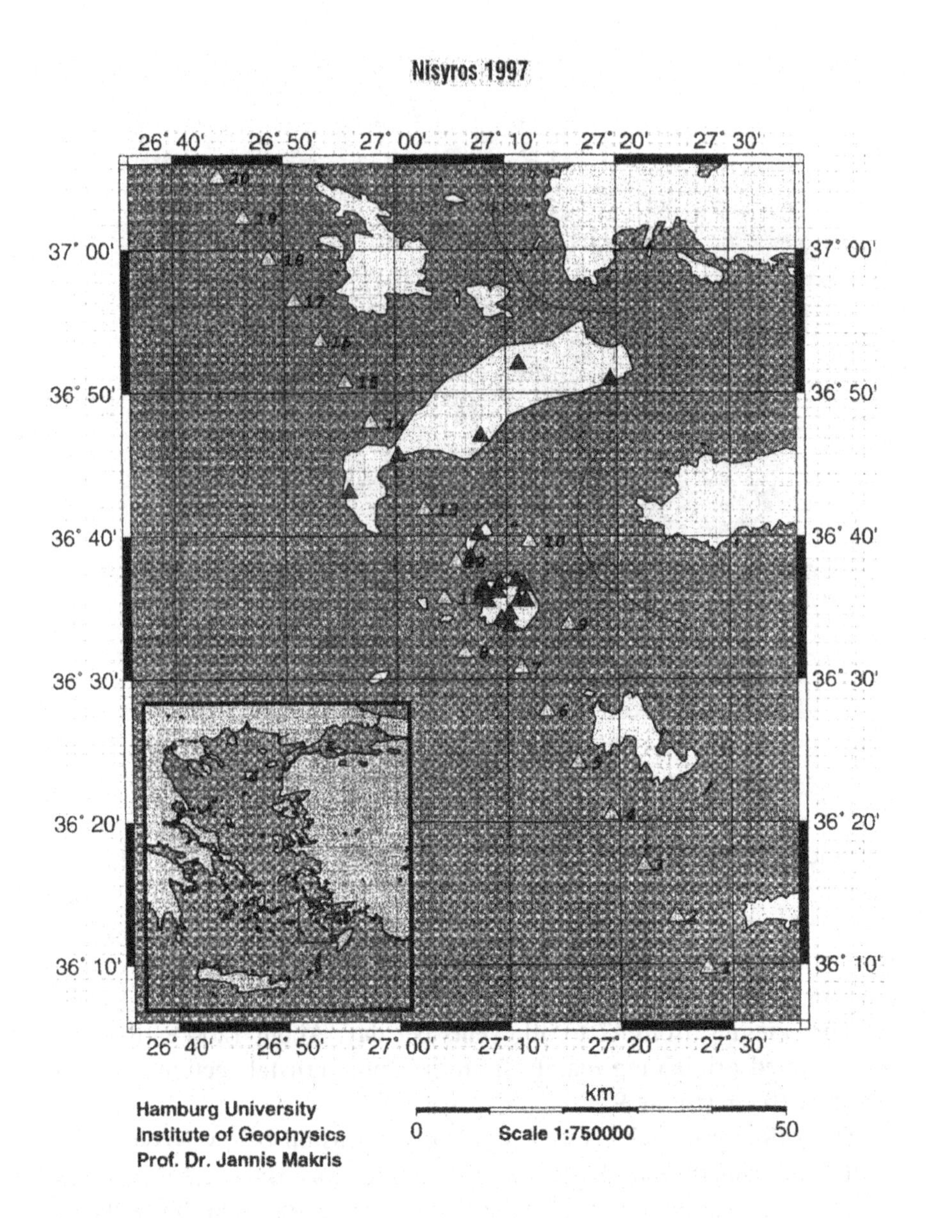

Figure 9: Microearthquake activity of Nisyros (Period of observation 11.10.97-29.10.97 and 03.11.97-23.11.97). The majority of the events are correlated with the caldera of Nisyros volcano. From Makris et al. 1998.

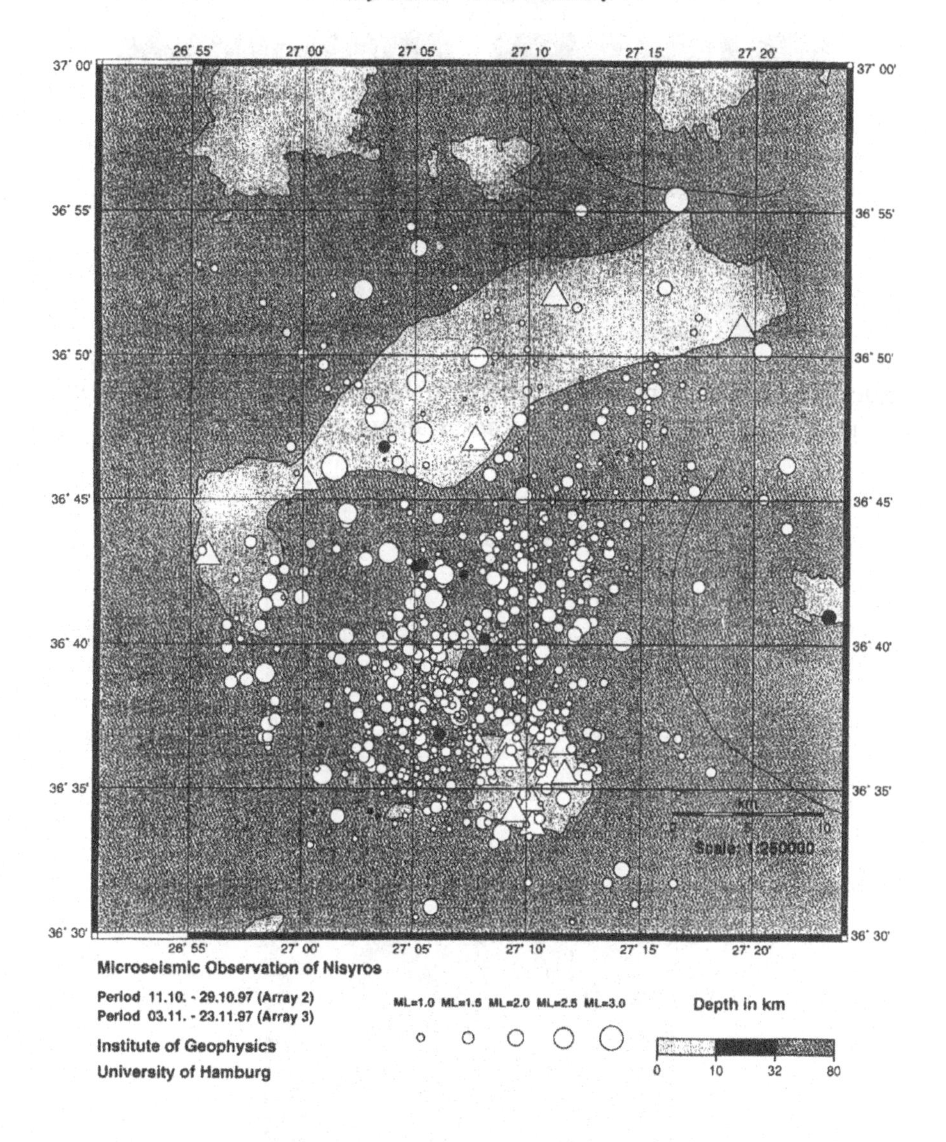

Figure 10: Nisyros area velocity structure. From Makris et al. 1998.

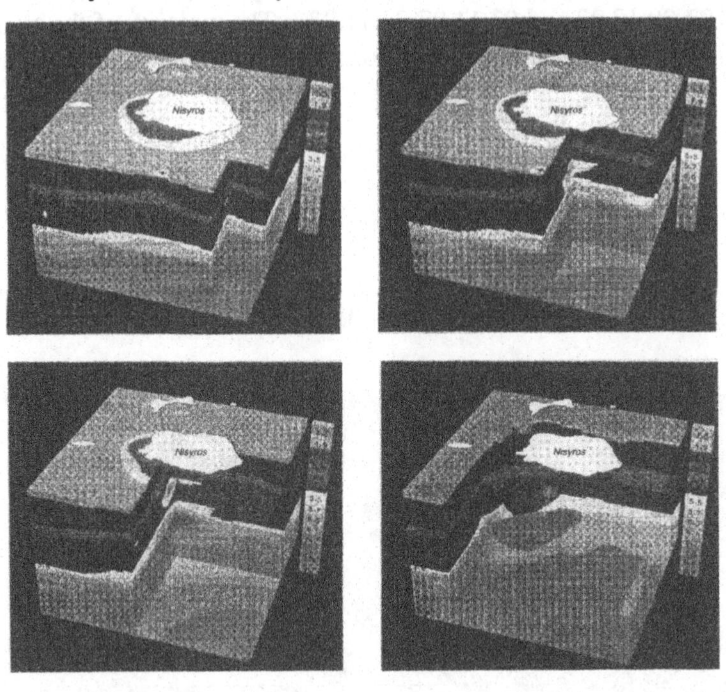

Figure 11: Location map of onshore/offshore seismograph stations in Messinia. From Institute of Geophysics, Hamburg

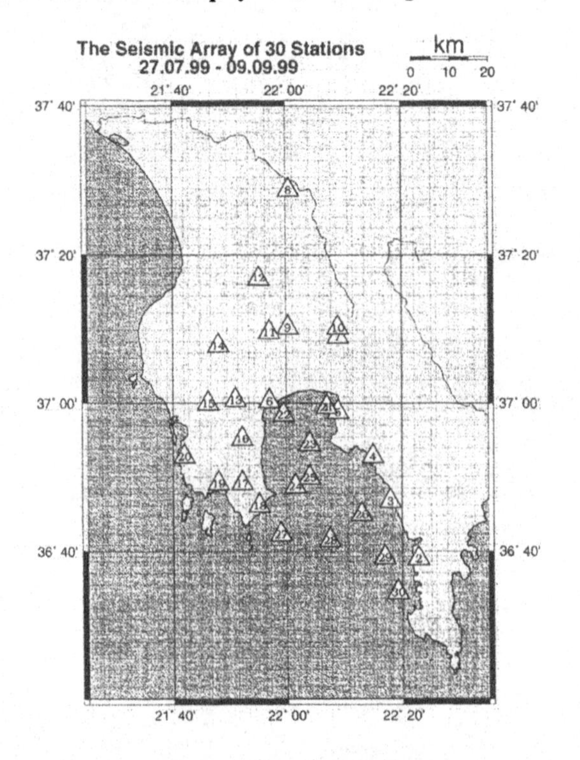

In the summer of 1999, a microseismicity survey was carried out in the inner Messiniakos gulf and surrounding region of Peloponnesus, southwestern Greece. The area is one of the most active regions in the Hellenic Arc, and was the host of the 1986 severe earthquake (Ms=6.2) that caused extensive damages and 20 human casualties in the urban center of Kalamata. 20 digital land stations and 10 ocean bottom seismographs, covering an area of approximately 3,500 square meters (figure 11), were deployed and operated for a two month period. The preliminary analysis of the seismic records showed a high microseismic activity originating mainly from the submarine part of the Messiniakos gulf. The evaluation of the results of this survey will be completed in early 2000.

3. CONCLUSIONS

During the last decade, seismological research has been extensively conducted in the Aegean and surrounding region. In this respect, the Institute of Oceanography of the National Center for Marine Research and the Institute of Geophysics of the University of Hamburg have for the first time initiated the activities in marine seismological research, by planning a series of projects for the investigation of the crustal type and structure in the submarine part of the Aegean, the identification of active fault zones, and the detailed and accurate recording of seismicity offshore.

During these projects, an enormous number of seismic and microseismicity data have been gathered, the analysis of which significantly contributed to the knowledge of seismicity and interpretation of the geodynamic processes in the Aegean and surrounding region.

It is worthy to notice, that the capacity of the Permanent Seismograph Network of Greece presents a significant gap in the southern part of the Aegean, along the Hellenic arc, due to its geometry. This particular region has been the host of many severe earthquakes that have caused many catastrophes in the Greek territory (Papazachos and Papazachou, 1989). The use of ocean bottom seismographs, improves the geometry of the seismograph networks, and assures the highest possible precision in earthquake epicenter determination.

The detection of small magnitude events helps to identify any anomalous variations in seismic energy release. Finally, the identification of submarine neotectonic faults, and their correlation with active fault zones onshore significantly contributes to seismic hazard and microzonation studies, as well as to earthquake risk mitigation in the presently active zone of the Aegean and surrounding areas.

REFERENCES

Bohnhoff, M., Makris, J., Papanikolaou, D. Stavrakakis, G. 2000. "Crustal investigation of the Hellenic subduction zone using wide aperture seismic data". *Tectonophysics* (submitted).

Dewey, J. and Sengor, A. 1979. "Aegean and surrounding regions: complex multiplate and continuum tectonics in a convergent zone." *Bull. Geol. Soc. Am.* 90: 84-92.

Galanopoulos, A. 1955. "Seismic Geography of Greece." *Ann. Geologiques des Pays Helleniques* 6: 83-121.

Le Pichon, X. and Angelier, J. 1979. "The Hellenic arc and trench system: a key to the neotectonic evolution of the eastern Mediterranean." *Tectonophysics* 60: 1-42.

Le Pichon, X., Chamot-Rooke, N., Lallemant, S., Noomen, R. and Veis, G. 1995. "Geodetic determination of the kinematics of central Greece with respect to Europe: Implications for eastern Mediterranean tectonics." *J. Geophys. Res.* 100: 12675-12690.

Makris, J., Papanikolaou, D., Stavrakakis, G., Liu, X., Papoulia, J., Sachpazi, M., Sakellariou, D. and Wodtke, D. 1996. "Detection of thin continental crust in the N. Evoikos gulf, central Greece." XXV General Assembly of the European Seismological Commission, Reykjavik, Iceland (Abstracts).

Makris, J., Chonia, T., Papanikolaou, D. and Stavrakakis, G. 1998. "Active and passive seismic Studies of Nisyros Volcano – East Aegean Sea." 3rd International Conference on the Geology of East Mediterranean, Cyprus.

Makris, J., Papoulia J., Papanikolaou, D., Stavrakakis, G. 2000: "Thinned continental crust below northern Evoikos basin, central Greece, detected from deep seismic soundings". *Tectonophysics* (submitted).

Makropoulos, K., Drakopoulos, J. and Latoussakis, J. 1989. "A revised and extended earthquake catalogue for Greece since 1900." *Geophys. J. Int.* 98: 391-394.

McKenzie, D. P. 1970. "Plate tectonics of the Mediterranean region." *Nature* 226: 239-243.

McKenzie, D. P. 1972. "Active tectonics of the Mediterranean region." *Geophys. J.R. astr. Soc* 30: 109-185.

McKenzie, D. P. 1978. "Active tectonics of the Alpine-Himalayan belt: the Aegean sea and surrounding regions (tectonics of Aegean region)." *Geophys. J.R. astr. Soc.* 55: 217-254.

Papazachos, B. and Comninakis, P. 1971. "Geophysical and tectonic features of the Aegean arc." J. Geophys. Res. 76: 8517-8533.

Papazachos, B. and Papazachou, C. 1989. "The earthquakes of Greece." Ziti Publications, Thessaloniki, 304pp.

Papazachos, B. 1990. "Seismicity of the Aegean and surrounding area." *Tectonophysics* 178: 287-308.

Ryan, W., Hsu, K. and Stanley, D. 1973. Initial reports of the Deep Sea Drilling Project, 13, 1.2, U.S. Government Printing Office, Washington D.C.

Taymaz, T., Jackson, J. and McKenzie., D. 1991. "Active tectonics of the north and central Aegean Sea." *Geophys. J. Int.* 106: 433-490.

4

Seismic Risk in Romania:
Features and Countermeasures

Horea Sandi
INCERC
Building Research Institute
Bucharest, Romania

1. INTRODUCTION

Seismicity is widely recognized to represent the most severe natural hazard in Romania. The record of the current century ascertains this belief. Romania underwent in this century two destructive earthquakes, both of them generated by the Vrancea intermediate depth seismogenic zone, on 1940.1.10 (M = 7.4) and on 1977.03.04 (M = 7.2) (where M represents Gutenberg - Richter magnitudes) respectively. The toll taken by the 1977 event, which was the most destructive earthquake of modem history of Romania, was represented, according to official data published at the end of 1977 [1], by:

- 1,570 lives lost and 11,300 persons injured;
- 32,900 apartments destroyed or heavily damaged;
- 53,000 homeless families;
- 763 economic units heavily affected;
- total economic losses exceeding US$ 2 billion.

The 1977 event was followed by the Vrancea earthquakes of 1986.08.30 (M = 7.0), 1990.05.30 (M = 6.7) and 1990.05.31 (M = 6.1) and by a sequence of (crustal) Banat earthquakes having occurred during the second half of 1991, out of which the strongest were those of 1991.07.12 (M = 5.7), 1991.07.18 (M = 5.6) and 1991.12.02 (M = 5.5).

While up to the 1940 earthquake there was no awareness about seismic risk in the community of engineers, the concern for earthquake protection

P.R. Kleindorfer and M.R. Sertel (eds.), Mitigation and Financing of Seismic Risks, 53–69.

became increasingly strong during the post-war period. The first official earthquake resistant design code of wide use was endorsed in 1963. The community of engineers was sufficiently qualified in 1977 to learn from that destructive earthquake. This awareness **is** still increasing. It has become obvious for many that prevention of earthquake disasters requires the development of a comprehensive defensive strategy, to encompass traditional engineering activities (first of all engineering design), as well as numerous other highly important aspects, related to the control and reduction of all components of seismic risk affecting society.

This paper provides a brief introduction to some main factors influencing seismic risk, as well as of some main measures aimed to mitigate that risk (including earthquake resistant design and some complementary measures).

2. SOME SPECIFIC ASPECTS OF SEISMIC RISK IN ROMANIA

2.1 General

The aspects dealt with in this section are related to the seismic conditions, to the seismic vulnerability of buildings and of some other categories of elements at risk, and to the seismic risk affecting the same. This is a background for the developments of the next section, where some important necessary measures, whether they were already adopted or not, are dealt with.

2.2 Some data on the seismic conditions

The territory of Romania is affected by the activity of several seismogenic zones. The seismic hazard is described in simplified terms by the standardized map of zonation of [15]. This map is expressed in terms of *MSK* intensities, where the indices refer to return periods assessed as follows: "*1*" means about 50 years, while "*2*" means 100 years or more. The zonation maps used in engineering design [16] are different: a first map is expressed in terms of the basic acceleration coefficient k_s, while a second map is expressed m terms of the corner periods T_c of design spectra. The maps referred to are reproduced in Figures 1, 2, 3. The two zonation systems are quite equivalent, in case one considers the developments of [12].

Figure 1. Seismic zonation of Romania according to STAS 11100/1-93

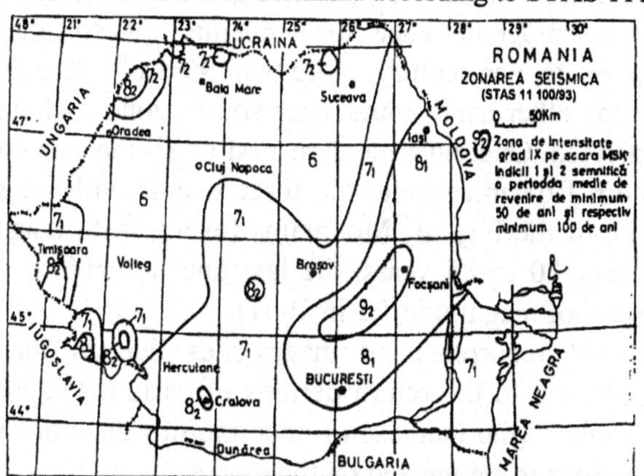

Figure 2. Zonation of Romania according to P.100-92
(as related to basic design factor k_s)

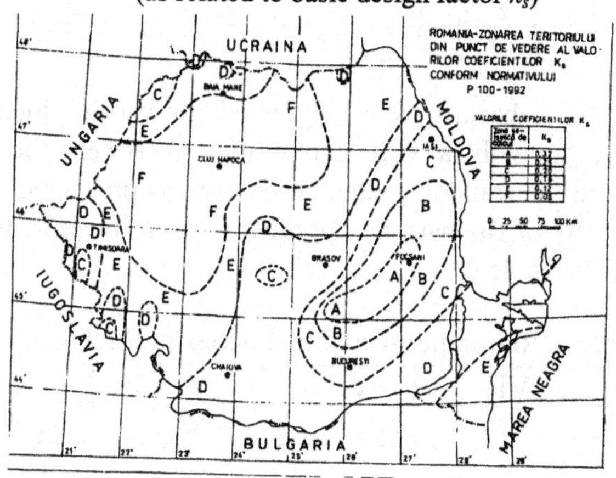

Fig. 3. Zonation of Romania according to P.100-92
(as related to corner period of dynamic factor T_c)

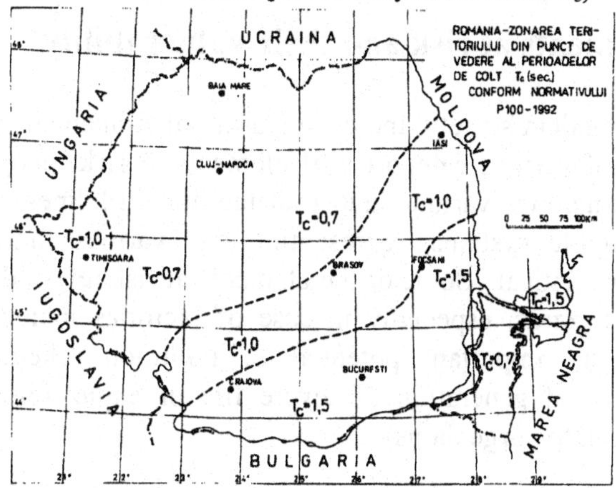

The Vrancea, intermediate depth, seismogenic zone [*VSZ*] is by far the most important seismogenic zone of Romania. According to [11], it delivers in the average, per century, more than 95% of the energy delivered in Romania. On the other hand, some other source zones, which are crustal, shall not be neglected. One must mention first of all the seismogenic zones of Banat (Western Romania) (note the three events with magnitudes 5.7, 5.6, 5.5 in 1991), of the F`g`ra] Mountains (events with magnitudes 6. to 6.5 occurring every 80 to 85 years, the last one in 1916) or of Southern Dobrogea (an event of magnitude 7.2 in 1901).

The analysis of recurrence of magnitudes due to the *VSZ*, [12], somewhat updated, led to the return periods of various Gutenberg-Richter magnitudes of Table 1 (the increase of uncertainties of return periods with increasing magnitudes is recognized by the increasing intervals given).

Table 1. Return periods of Vrancea magnitudes

M	7.4	6.	6.5	7.	7.2	7.5	7.6
T_{ret} (yrs.)	80...100	6	14	32	45...50	120...190	250...650

Estimates on local hazard were conducted for various localities using Cornell's approach, with a full convolution between the magnitude recurrence law and the (random) attenuation law, as presented in [12]. The outcome for the City of Bucharest, expressed in terms of *MSK* intensities is approximately as in Table 2.

Table 2. Return periods of intensities in Bucharest

I	VI	VII	VIII	IX
T_{ret} (yrs.)	10	20	50	200

The data of Table 2, for intensities $I \leq$ VIII., correspond quite well to the experience of the current century.

2.3 Elements at risk and their vulnerability

As in any modem society, the examination of Romanian society reveals the existence of various categories of elements at risk: people, buildings, engineering structures, various industrial facilities, lifelines, networks etc., even less tangible systems, representing e.g. various functions, among which some are critical. Some of the elements at risk referred to are related to secondary hazards, especially in case of facilities including high risk sources with an important potential of pollution (chemical, nuclear, bacteriological), of generation of severe fire or explosions, or of heavy flooding (especially large dams).

The seismic vulnerability of the various categories of elements at risk is known at different degrees of accuracy and certainty, depending upon their nature. It may be stated that the best degree of knowledge was achieved to date for the residential building stock. This is due to the experience of strong earthquakes, as well as to the development of methods of engineering analysis.

An extensive post-earthquake survey of the performance of buildings was undertaken subsequent to the 1977 earthquake. More than 18,000 buildings were investigated in Bucharest [2]. The results on their vulnerability, detailed for eight classes of buildings, were presented in [7]. Based on these results, on results from literature and on regression analysis, some analytical expressions were proposed for the vulnerability of buildings [11].

Considering now the case of residential buildings, which represent, of course, a category of elements at risk of highest importance, one may emphasize the considerable difference that exists between various categories of such buildings. Their vulnerability depends upon the period of construction (which determines the design philosophy), the materials, the type of loadbearing structure, as well as the height. The experience of the 1977 earthquake showed that the most vulnerable category is that of relatively tall buildings (about 8 to 12 storeys tall) built during the pre-war period. The main causes of their high vulnerability are:

- the lack of concern of design engineers for earthquake protection;
- the often poor material quality;
- the cumulative negative effects of the successive earthquakes referred to in the introduction;
- the effects of corrosion, sometimes of fatigue due to urban traffic too;
- inappropriate interventions undertaken sometimes by occupants;
- the tendency to resonance, during strong earthquakes.

28 out of the 32 buildings having collapsed in 1977 in Bucharest pertained to this category. There is a consensus that in case a similar earthquake strikes again, more cases of collapse are bound to occur.

This category of buildings is not the single one to be more vulnerable than acceptable. Several categories of residential buildings, designed according to the code in force during the pre-1977 period do not meet the requirements of the code currently in force [16] and this is strongly correlated with the data of post earthquake surveys, which put to evidence damage, sometimes heavy, inflicted by earthquake action.

There are also other elements at risk, which appear to be highly vulnerable. There are several industrial facilities, some of them including high risk sources (especially related to pollution of various kinds) that are not in a satisfactory condition. Corrosion due to the particular environment that can be met at various places in industry is one reason. Other reasons may be related to fatigue, to overloading due to malfunctioning of equipment (cranes, machines etc.), to poor quality of construction works etc.

2.4 Implications for seismic risk

As is well known, seismic risk results as a consequence of seismic hazard, seismic vulnerability and exposure of various elements at risk. It is hard to try a global, even rough, estimate of seismic risk affecting Romania, but it may be mentioned that a consensus exists among professionals about the likely impact of a future strong earthquake: in case an earthquake like that of 1977 strikes again soon, its impact would be more severe than that witnessed in 1977. Some main reasons for this situation are as follows:

- several categories of residential buildings are in a poorer condition than on the eve of the 1977 event (main reasons: the cumulative effects of the strong earthquakes underwent meanwhile and of corrosion).,
- the ability of the state to control and concentrate various kinds of resources in order to react under emergency conditions is lower now.

An attempt to analyze risk affecting existing buildings was made in [11]. Probabilistic tools were used and a parametric analysis was performed for seismic conditions like those of Bucharest. Alternative assumptions were considered for the nominal level of earthquake protection and for the duration of exposure. The model used for the stochastic processes related to hazard and risk was Poissonian. The outcome of these analyses showed among other that, in case pre-war tall buildings are left as they are, without undertaking measures to reduce their vulnerability, for decades to come, the risk of collapse goes into the range of tens of percents. This conclusion, since it is in quite good agreement with the experience of the 1977 earthquake, when the 28 buildings having collapsed represented 7% of a stock of about 400 buildings of this category, **is** of course most alarming.

The possibilities to assess risk affecting other categories of elements at risk are more modest, first of all due to the very limited information on their inventory and on their vulnerability. A main source of information in this respect is given by the experience of the 1977 earthquake. This experience shows, among other, that chains of events did not play a major role in relation to the physical damage to various elements at risk. On the other hand, it is likely that indirect economic effects were considerably more severe than what would correspond to the official estimates of somewhat more than 2 * `10^9 US$, published at the end of 1977. The balance of foreign trade was seriously upset after 1977 and this might be easily due to an important extent to the decrease of exporting capabilities following the earthquake impact.

3. SOME REFERENCES TO THE REGULATORY BASIS OF DESIGN

The regulatory basis related to earthquake protection consists of two main components:

- standards and codes used in engineering design, aimed to provide the required earthquake resistance to various artefacts of man;
- complementary regulations, aimed to promote measures for earthquake risk control and reduction.

Following data are related to the development of the existing regulatory basis of design.

Following the 1940 earthquake, the Ministry of Public Works issued instructions related to the earthquake protection of public structures in 1942 and then in 1945. Those documents were not widely applied in engineering practice.

A first official standard on seismic zonation was issued in 1952, as STAS 2923-52, and was revised as STAS 2923-63, STAS 11100/1-77, STAS 11100/1-91 and SR 11100/1-93. This standard expresses seismic conditions in terms of *MSK* intensities. Those intensities were intended at the beginning to mean maximum observed intensities, while since 1991 they represent intensities corresponding to some explicit return periods. Note here that zonation is expressed, in a quite equivalent manner, also in terms of engineering design parameters, by the code for design referred to subsequently.

A first official code for earthquake resistant design of residential, public and industrial structures was issued in 1963, as P.13-63, and revised as P.13-70. After the 1977 earthquake, this was changed to P.100-78 and then P.100-81. The experience of the 1986 and 1990 earthquakes was at the basis of the versions P.100-91 and P.100-92. The last two sections of this latter version, which are concerned with the existing building stock, were revised in 1996. Both P.13 versions relied on foreign experience and on the provisions of the Soviet code of the time, which relied, at its turn, also on the Californian code.

The 1978 version brought considerable improvements. Besides an improved system of design parameters, it introduced for the first time some quantitative criteria intended to ensure ductile structural solutions. Continued progress in the same sense characterized the 1991 and 1992 versions. Important additional developments were introduced then with the two sections concerning the evaluation of existing structures and their strengthening. Those latter sections were revised, as mentioned, in 1996, based on a summary of experience of their application in practice.

A new period was recently initiated, in the sense of passage to a new regulatory basis, compatible with the Eurocodes.

4. PROBLEMS RAISED BY THE HARMONIZATION WITH EUROCODES

4.1 General

There exists a consensus in the professional community of civil engineers of Romania that the regulatory basis of structural design must be modernized and that the system of Eurocodes must be taken as a reference in this connection. There are two main reasons of this consensus:

1. The political option for European integration, which meets a wide consensus of the population.
2. The recognition of the high scientific and technical level of the system of Eurocode drafts, which represents a model to be followed in order to improve the quality of the Romanian regulatory basis.

Although there does not formally exist to date in Romania a detailed, firm, comprehensive strategy in this view, several convergent steps were undertaken in order to promote at least a harmonization with the system of Eurocodes, but perhaps even some more advanced steps towards integration. A system of *CR*'s (Romanian codes) compatible with the Eurocodes was partially planned to be developed. In relation to the specific field covered by *EC*-8, it may be mentioned that three draft codes for bridges [3] (with a quite advanced harmonization with *EC*-8 / Part 2), for tunnels and galleries and for retaining walls were developed, and that a volume on worked examples on reinforced concrete structures, on steel structures and geotechnical aspects preceded by information on the representation of seismic action [5] was edited too. Drafts corresponding to *EC* -8 / Parts 1-1, 1-2, 1-3 are also on the way to be developed.

The current stage of the Romanian regulatory basis in the field of earthquake resistant design and the current trends, briefly referred to above, raise several problems like:

- identification of the needs of structuring the regulatory basis of earthquake resistant design, in order to cover all needs of practical activity;
- option on the conceptual basis in this field;

- calibration of the parameters influencing the level of earthquake protection;
- option on the relationship with the regulatory basis developed at European level.

The trends referred to are related to a period of profound transition in Romania. A similar transition period occurred in the past in relation to the adoption of the limit states method. Following a decision taken in 1960, in connection with the adoption of the limit state design method, using the Soviet regulatory basis in force since 1955 as a model, two main steps had to be undertaken. A first step, completed in 1962-1963, led to the enforcement of a system of "conditional" codes (applied rather experimentally) concerning:

- the basis of structural design
- actions on structures;
- design of plain, reinforced and prestressed concrete members;
- design of steel members;
- design of masonry structures;
- earthquake resistant design.

A first version of a standard for the design of plain, reinforced and prestressed concrete members on the basis of the limit state method, that developed additionally the conditional code referred to, was endorsed in 1969. The system of conditional codes was converted to an unproved system of standards which were endorsed in 1975-1978 and were revised subsequently several times, without major changes. This system of standards is still in force at present.

It may be noted that the main modifications adopted after the period of enforcement of the standards were due to the occurrence of the impact of the 1977 earthquake. Provisions related to earthquake protection were included in some regulatory documents, like a standard on classification and combination of loads, as well as in the standard on design of concrete members. Important modifications were brought to documents like the earthquake resistant design code, the codes for design of masonry structures, for design of buildings with reinforced concrete structural walls, or stimulated the development of a code for reinforced concrete frame structures.

The periodization related to the introduction of the limit state method means relevant experience in relation to the efforts required by the passage to a new generation of codes. Some difficulties that must be surpassed in future in connection with the harmonization with the system of Eurocodes (which is itself in a stage of drafts to date) are related to following aspects:

- the change involved by a passage to a regulatory basis harmonized with the Eurocodes is deeper than the changes involved by the goals of the two previous periods referred to;
- the modification of requirements of professional qualification for the community of users of a new regulatory basis is more important than in the past periods;
- the groups involved in code drafting are older than in the past;
- the period of political and economic transition involved by the historical changes initiated in 1989 led to tendencies of atomization of the professional community.

4.2 Elements of comparison between Romanian code and EC-8 and related suggestions

4.2.1 General

A comparison of the system of Romanian national codes devoted to earthquake resistant design with the current stage of development of the *EC*-8 puts to evidence differences from the viewpoints of scope and structure of the system, of the conceptual basis and of the nominal earthquake protection level. Some suggestions on how to further develop the Romanian national system start from the following premises:

- the structure of *EC*-8 (with limited improvement/completion) should be taken as a starting point;
- the structure of the system should be extended according to practical needs, keeping such codes as those related to the design of shear wall or frame structures and developing new documents where this would be useful for engineering practice;
- harmonization should be based on adoption of symbols and definitions of *EC*-8, as well as of the system of presentation of principles, requirements and criteria;
- the current safety level provided by the Romanian regulations should be brought closer to what corresponds to the level of [475] years, specified by *EC*-8.

Further aspects dealt with in this section are related to some problems of general interest, without discussing aspects that are specific to some categories of structures. One should consider in this connection also the suggestions of Section 5 in relation to possible modifications and further developments of *EC*-8.

4.2.2 Basic design parameters

Three basic representations of seismic action are considered in *EC*-8:

- *R.Sp.* (design spectra)-,
- *R.Ac.* (design accelerograms),
- *R.St.* (stochastic representations).

Accepting that, at the current stage of development, the representation *R.Sp.,* based on specification of response spectra, is the reference representation, the problems related to the format and calibration of specific parameters are as follows. The system adopted in *EC*-8, consisting of specification of *elastic response spectrum* (relations (4.1)..(4.4) of Part. 1. 1) and of *design response spectrum for linear analysis* (relations (4.7) ... (4.10) of Part 1) is analytical and also instructive for users. It may be suggested for adoption, even if it is not clear why it was necessary to adopt different physical dimensions for the two entities. (LT^{-2} for the first, *1* for the second). The response spectrum format is, of course, not sufficient in order to cover various specific situations occurring in design. It should be completed in a systematic way with specifications on how to consider *NDOF* input, corresponding to spatial ground motion, as e.g. in Part 2 of *EC*-8 for bridges, in Part 3 of *EC*-8 for towers, and in Annex C of P.100-92 for buildings. This aspect is dealt with again in Subsection 4.3.

The problems of acceptability of the calibration of the reference parameters a_g (T_B, T_C, T_D), S and q deserve nevertheless some discussion. The return period of [475] years is much longer than that of 50 years accepted at present and, according to the results of [12], the adoption of the increased return period would at least double the values used in current practice. Without denying the importance of considering increased return periods and corresponding values a_g, this problem should be subject of further research and debate. Some attempts to verify existing buildings for a_g = 0.35 g instead of a_g = 0.2 g, as specified for Bucharest by the code P.100-92, showed that this is feasible in case one uses increased material strength values, corresponding to statistical means. The increased a_g value corresponds to a return period of some 200 years, which represents a considerable improvement as compared with 50 years accepted at present. The extended use *of* a_g values of such a level needs nevertheless adaptation from several viewpoints: after some more systematic experimental use, revision of criteria of verification and, last not least, acceptance by the professional community after some actions of information and education.

The system of values (T_B, T_C, T_D) must be reconsidered, in order to make it fit with what is known from our own records. The most important aspect is that, according to records at hand, the corner period T_C given in *EC*-8 is too short and this might be due, to an important extent, not only to local conditions, but also to the features of focal mechanisms. It may be also noted that the corner period referred to seems to increase considerably with increasing magnitudes, so that taking decisions on the basis of the 150 accelerograms recorded in 1986 and 1990, for earthquakes with Gutenberg - Richter magnitudes not greater than 7.0, could be hazardous. The system of values q given by *EC*-8 appears to be in general appropriate, but its acceptance should be conditional upon a verification of existence of possibilities of mobilization of the ductility resources of structural members.

The system of design forces, as specified by the relations (3.4), (3.5) of Part 1.2 of *EC*-8 should be presented in more complete form, as given in Section 5 and Annex C of P.100-92, considering also the generalizations required for NDOF input. This system should be accompanied by specification of data for design of equipment, piping etc. installed on structures.

4.2.3 Alternative representations of seismic action

The representation *R.Ac.*, considered in *EC*-8, as well as in P.100-92, should be kept in future. It is necessary, nevertheless, to specify in more detail the rules of selection of design accelerograms in connection with data on site conditions and, what is perhaps more important, to explicitly consider the alternative possible objectives of engineering analyses:

- O_1: numerical experiment for an individual accelerogram;
- O_2: analysis of sensitivity of response with respect to the variation of some input parameters,
- O_3: vulnerability analysis (based on a Monte-Carlo approach).,
- O_4: full (probabilistic) risk analysis.

The representation *R.St.*, considered explicitly in *EC*-8, but not in P.100-92, should also be referred to explicitly in future. Emphasis should be put on its fundamental role for the structure of basic relations.

4.2.4 Specification of requirements and criteria

The consideration of a system of two categories of limit states (ultimate LS's and serviceability LS's) is common m *EC*-8 and in P.100-92, with the difference that the concept of serviceability limit states is more explicitly considered in *EC*-8. On the other hand, the experience of verification of existing structures revealed the need to consider a more complete mobilization of resistance reserves for structures, which do not fulfil the criteria related to the ultimate limit state. In fact, the ultimate limit state formally considered in *EC*-8 and P. 100-92 as well, involves neglecting of considerable resistance reserves provided by overstrength, as well as by the mobilization of resistance reserves due to nominally non-structural components. In order to have a more complete picture on safety and reliability, if would be appropriate to consider three categories of limit states: serviceability LS, (nominally) ultimate LS and collapse LS. Engineering verifications should be conducted for design loads corresponding to return periods differentiated with respect to the nature of limit states and to the classes of importance of structures too. This differentiation should replace the consideration of the importance factor γ_I.

According to the belief of the author, the Eurocode *EC*-8 represents a model of first importance for the development of the homologous regulatory basis in Romania and it may be confidently stated that the way of harmonization will be followed. On the other hand, one may consider some needs of refinement of the regulatory basis also beyond the stage reached in the 1994 *EC*-8 draft. The author presented some suggestions on this subject in [9]. Without going into more developed analytical considerations, one may state that more consistency should be required for the calibration of design parameters, with appropriate differentiation for different classes of importance and of different limit states. It is reasonable to consider in this connection three categories of limit states:

- *LS. 1*: serviceability,
- *LS. 2*: nominal ultimate (in the sense of conventional non-exceedance of the elastic stage);
- *LS. 3*: collapse (total or partial).

A systematic way to specify the basic design parameter related to the reference ground acceleration is to prescribe values corresponding to different return periods. The calibration of Table 3 may be proposed for discussion.

Table 3. Proposed return periods for basic design parameter

Class of importance	Return period (years) for the limit states		
	serviceability	(nominal) ultimate	collapse
I	50	200	1000
II	20	100	500
III	10	50	200
IV	5	20	100

5. SOME REFERENCES TO RISK MANAGEMENT

5.1 General

The high level of seismic risk estimated for Romania requires a systematic, wide scale, concern for control and mitigation. Given the complexity of tasks involved, a comprehensive strategy must be developed in this connection. This strategy must deal with

- specification of rules for the protection of new developments at the level of individual buildings, of industrial facilities, of infrastructure, of medical network etc.;
- specification of a policy on how to deal with the existing elements at risk, in order to control and mitigate seismic risk.

Experience shows that attempts to deal in this sense with existing works encounter often very high difficulties, due to reasons of technical, financial or social nature. So, this section of the strategy will require the greatest efforts.

Without attempting at this place to discuss more general aspects of risk management, it may be of interest to emphasize some urgent measures to be considered, primarily by the Ministry of Public Works and Territorial Planning.

5.2 Some urgent measures proposed

Some measures proposed to be urgently adopted are presented further on. It may be stated that the adoption and implementation of these measures will require neither important costs, nor a long time and that their adoption will be most beneficial from the viewpoint of earthquake protection goals.

1. A program of inventory, preliminary classification and evaluation of elements at risk, in agreement with the tasks foreseen for every holder or caretaker of various elements at risk. A draft Government decision to update and adapt the provisions of some previous decisions for various sectors of activity (this should include the obligation to develop a network of databases for the results of inventory, classification and other related activities).

2. Preparation and urgent endorsement of regulations aimed to urge evaluation activities and, if necessary, proper intervention, upon public buildings held by central agencies, upon essential facilities (e.g., some medicare facilities, firemen units, schools etc.), upon facilities including high risk sources etc., ensuring also appropriate financial mechanisms for this purpose.

3. Urgent identification of residential buildings affected by a high risk of collapse in case of incidence of an earthquake similar to those of 1940 and of 1977, and effective intervention upon them within a tome interval of 3-4 years.
4. A program of putting out of operation (temporarily or definitively) of industrial facilities affected by a high risk, due to the current state of corrosion, upon which the holder cannot urgently intervene.
5. A program of evaluation and if necessary of design of solutions of intervention upon repetitive structures proven to be unsafe.
6. The development of well correlated criteria of urban development and of intervention upon the existing building stock in connection with the general rehabilitation and upgrading goals, in order to achieve an urban development under conditions of a satisfactory earthquake protection level.
7. The development of an insurance strategy aimed to stimulate undertaking of earthquake protection measures by the holders of buildings and other structures.
8. A program of earthquake preparedness aimed to ensure prompt and efficient reaction in the event of a strong earthquake (investigation of affected structures, establishing intervention priorities, provisional earthquake risk reduction measures).
9. Specification of prerogatives and tasks of central agencies, in agreement with the provisions of the Government ordinance no. 47/1994, concerning the earthquake disaster prevention: categories of works to be surveyed (besides residential buildings, public buildings, industrial facilities. bridges, dams, nuclear facilities etc.), actions to be undertaken or surveyed, providing of necessary resources and of an appropriate legislative frame.
10. Development of detailed earthquake protection strategies for various sectors of activity, with participation of specific groups and of groups or offices aimed to implement these strategies.
11. Examination of the current stage of research and development of a program aimed to cover the current high priority needs.
12. Examination of the current stage of development of the regulatory basis and development of a program aimed to cover the current high priority needs.

The urgent measures enumerated do not conflict in any way with the development of a comprehensive and consistent earthquake protection strategy. On the contrary, they must be integrated into such a strategy, which must be considered itself an urgent task of high priority.

6. CLOSING CONSIDERATIONS

The short presentation provided in previous sections was aimed to offer a look at some problems related to seismic risk in Romania. Topics related to technical aspects of engineering activities were dealt with together with topics of more general, complementary, nature. A presentation of this length could not offer, of course, an in-depth insight into all aspects related to seismic risk and of efforts required for its control and mitigation.

The Romanian school of earthquake engineering, which developed gradually during the second half of the current century, made obvious its capacity on the occasion of the 1977 earthquake, when structures designed to resist earthquakes resisted better than in many other countries for an earthquake of this size. The Romanian school of earthquake engineering also demonstrated its ability to learn from the experience of that event and subsequent ones. This does in no way mean that continued progress is not needed and that there are not fields for which present skills must be considerably improved in order to face the challenge of risk reduction. There is a need of continued and extended education and of continued and extended research and this should cover all fields that are relevant for earthquake protection. There is also stringent need for databases of various categories, ranging from research topics to inventory of various elements at risk, to the outcome of their evaluation. There is also a need to develop functional structures able to tackle the risk reduction tasks and the emergency tasks of post-earthquake situations.

REFERENCES

B`lan, T., V. Cristescu, and I. Cornea (coordinators). 1982. *The Romania Earthquake of'4 March* 1977 (in Romanian). Editura Academiei, Bucure]ti, 1982.

Constantinescu, L., and V. Marza. 1980. A computer-compiled and computer-oriented catalogue of Romania's earthquakes during a millennium (984-1979). *Rev. Roum. de Géologie, Géophysique et Géographie* 24(2) .

Fierbineanu, V., and D. Lungu. 1998. The new aseismic design code for bridges in Romania (to be published). *Proc. 11-th European Conference on Earthquake Engineering,* Paris.

Georgescu, E. S., P. Popescu, H. Sandi, and O. Stancu. 1998. Towards a national earthquake protection program. In *Proc. International Workshop on Vrancea Earthquakes*, edited by F. Wenzel and D. Lungu. Dordrecht: Kluwer Academic Publishers.

Sandi, H. 1996. Some needs and suggestions concerning the development of design regulations. Proc. *11 - th WCEE,* Acapulco, 1996.

Sandi, H. 1998. Problems raised by the adaptation of EC-8 to the conditions of Romania *Proc. 11-th European Conference on Earthquake Engineering,* Paris, 1998.

Sandi, H. 1997. "Earthquake risk analysis and management." In *Proc. International Workshop on Vrancea Earthquakes*, edited by F. Wenzel and D. Lungu. Dordrecht: Kluwer Academic Publishers.

Sandi, H., and L. Floricel. 1994. Analysis of seismic risk affecting the existing building stock. *Proc. 10-th European Conference on Earthquake Engineering,* Vienna.

Sandi, H., and O. Stancu. 1995. Parametric analysis of seismic hazard for intermediate depth earthquakes of Romania. *Proc. 5-th International Conf. on Seismic Zonation,* Nice.

Government of Romania: Ordinance no. 20/1994 on the mitigation of risk affecting the existing building stock (in Romanian). *Monitorul Oficial al Rom@niei, part* 1, no. 28.

Government of Romania: Ordinance no. 47/1994 on defense against disasters (in Romanian), *Monitorul Oficial al Rom@niei, part* 1, no. 242.

IRS (Romanian Standardization Institute): *Seismic Zonation of Romania* (in Romanian). SR 11 100/1-93.

MLPAT (Romanian Ministry of Public Works): Code for the Earthquake Resistant Design of Residential, Public and Industrial Structures (in Romanian) P. 100-92.

MITIGATION AND MITIGATION CHOICE

5

Public-Private Partnerships for Reducing Seismic Risk Losses*

Howard Kunreuther**
Center for Risk Management and Decision Processes
The Wharton School
University of Pennsylvania
Philadelphia, PA 19104

1. INTRODUCTION

The importance of public-private partnerships for disaster management has been stimulated by losses from catastrophes in the United States and other parts of the world. Hurricane Andrew which created damage to Miami and Dade/County, Florida in September 1992 and California's Northridge earthquake together cost the insurance industry (US$28 billion) and the government an additional US$17.6 billion. The Kocaeli earthquake in Turkey in August 1999, which caused over 17,000 confirmed fatalities with massive disruptions to the economy of Western

---

* Support from NSF Grant CMS 97-14401 to the University of Pennsylvania and from the Wharton Risk Management and Decision Processes Center Managing Catastrophic Risk project is gratefully acknowledged. Helpful comments were received from participants at the Wharton Managing Catastrophic Risk Sponsors Meeting in Bermuda on May 15-16, 2000. Special thanks to Applied Insurance Research, EQE and Risk Management Solutions for constructing exceedance probability curves for the model cities and for their help in analyzing the impact of parameterized indexed cat bonds on the losses to insurers. Jaideep Hebbar, with the assistance of Pooja Goyal, Patricia Grossi and Vikram Prasad, spent many hours analyzing the data from Oakland and helped construct the relevant tables and figures in the paper.

** Cecilia Yen Koo Professor of Decision Sciences and Public Policy, Chair of Department of Operations and Information Management, The Wharton School, University of Pennsylvania

P.R. Kleindorfer and M.R. Sertel (eds.), Mitigation and Financing of Seismic Risks, 73–99.

Turkey, has led to a recognition by the Turkish government, industry, and the public of the urgent need to develop and enforce better building standards (Wilczynski and Kalavakonda 2000).

This paper explores risk management strategies for reducing losses from natural disasters and providing financial resources to victims of these devastating events in both developed countries and emerging economies. More specifically, it will examine programs that involve the private sector such as insurance and capital market instruments (e.g. Act of God bonds) in combination with public sector programs such as regulations and standards (e.g. well-enforced building codes). The focus of attention will be on the earthquake problem but the concepts are relevant to other natural and technological disasters as well.

Figure 1 depicts a framework for analyzing this problem. It builds on concepts developed in a report by the Heinz Center (1999) and by Kleindorfer and Kunreuther (1999). The vulnerability of a city or region includes the potential for direct damage to residential, commercial and industrial property, other facilities such as schools, hospitals, and government buildings as well as infrastructure damage to highways, water, gas, electricity and other lifelines. Any disruption of infrastructure such as loss of the water supply or electric power can cause indirect losses by interrupting business activity, forcing families to evacuate their homes and causing emotional stress to families. One also has to consider the exposure of the population to the hazard and the potential number of fatalities and injuries to different socioeconomic groups.

The ingredients for evaluating the vulnerability of a city or region to natural hazards are risk assessment and societal conditions. Ideally a *risk assessment* specifies the probability of events of different intensities or magnitudes occurring and the impact of the direct and indirect impacts of these events to the affected interested parties. *Societal conditions* include human settlement patterns, the built environment, day-to-day activities and the institutions established to deal with natural hazards.

Before developing a disaster management strategy one needs to understand the decision processes of the key stakeholders. The term decision processes refers to the type of information and data collected by individuals, groups and organizations (either private or public) and how they are utilized in making choices. For example, if a family is considering whether to bolt its home to a foundation to reduce future losses from a severe earthquake, what information does it collect on both the hazard itself and the potential damage with and without this mitigation measure? What type of decision rule(s) does the family utilize in determining whether or not to invest in this mitigation measure? What type of data and decision rules do insurers, financial institutions and public sector agencies utilize in evaluating the cost-effectiveness of different mitigation measure? Unless we understand the nature of the decision processes of these different interested parties, we will have a difficult time recommending specific programs or policies.

Figure 1: Framework for Analysis

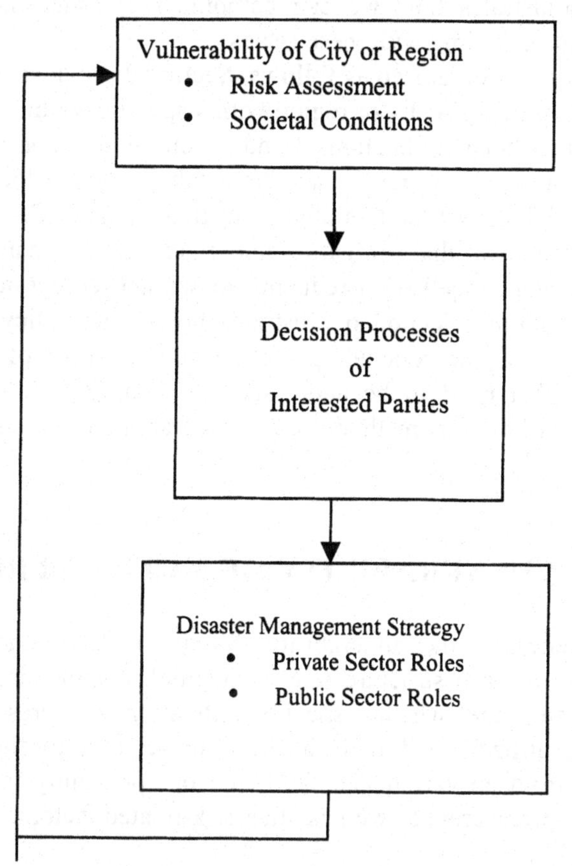

Based on an understanding of the vulnerability of the city or region and the decision processes of the key interested parties, one needs to develop a strategy for reducing losses and providing financial protection to victims of future disasters. This strategy will normally involve a combination of private and public sector initiatives which include insurance and new financial instruments as well as well-enforced building codes and land-use regulations. These measures will differ from country to country depending upon the current institutional arrangements and existing legislation and laws.

We apply this framework to Oakland, California, one of three cities that the Wharton Managing Catastrophic Risk project has been studying.[1] In particular, we examine the potential losses to either a private sector entity (e.g. an insurance company) or a public sector agency providing financial protection against losses from an earthquake in Oakland. This analysis illustrates how we can combine risk management tools with mitigation for dealing with these issues.

The paper is organized as follows: Section 2 provides a perspective on the vulnerability of a city or region with a specific application to Oakland, CA. We then focus in Sections 3 and 4 on the decision processes of the homeowners and insurers—two key stakeholders affected by natural disasters and show how this applies to the Oakland, CA case. Section 5 turns to the ways that mitigation and financial risk management policy tools can work together in reducing losses and addressing post-disaster needs of victims. Section 6 illustrates how these policy tools apply to Oakland, CA. The concluding section outlines a set of future research questions that need to be addressed for dealing with the problem by focusing on how this methodology can be applied to a study of Istanbul, Turkey.

2. VULNERABILITY OF A CITY OR REGION

In determining the vulnerability of a city or region one needs to know the design of each structure (e.g. residential, commercial, public sector) and infrastructure, whether specific mitigation measures are in place or could be utilized, and their location in relation to the hazard. (e.g., distance from an earthquake fault line or proximity to the coast in a hurricane-prone area) as well as other risk-related factors.

2.1 Constructing an Exceedance Probability Curve

Based on this information one can construct an exceedance probability curve, which depicts the annual probability that the losses from a series of different disasters will exceed a certain magnitude. The EP curve is the key element for evaluating a set of risk management tools. The accuracy of the EP curve depends upon the ability of scientific experts and

[1] The three cities, their associated modeling firm and hazard in parentheses are: Miami/Dade County, FLA (Applied Insurance Research,-hurricanes); Long Beach, CA (EQE, earthquakes); Oakland, CA (Risk Management Solutions, earthquakes).

engineers to estimate the impact of disasters of different magnitudes on different structures.

With respect to the earthquake hazard, scientists have been working to reduce the ambiguity and uncertainty in predicting the location, severity, frequency of occurrence, and physical effects of earthquakes by examining geologic records, looking at actual events, and conducting experiments on how the ground responds to earthquake processes. However, scientists are still uncertain as to how different factors interact with each other and their relative importance (Hanks and Cornell, 1994).

Engineers have focused on the nature, distribution, and level of damage from earthquakes. Such investigations have increased our understanding of the performance of various types of buildings and structures in earthquakes of different magnitudes. Hazard risk maps have been drawn for earthquakes, but they only provide rough guidelines as to the likelihood and potential damage from specific events.[2] The recent use of geographic information systems (GIS) for incorporating geologic and structural information for a region has enabled scientists to estimate potential damage and losses from different earthquake scenarios. The data for the region are stored in the form of GIS maps of ground shaking estimation, maps secondary seismic hazards such as liquefaction, landslide and fault rupture and maps of damage to structures in the region. (King and Kiremidjian in press).

2.2 Application to Oakland

We illustrate these concepts by developing an EP curve for a hypothetical insurance company that is providing financial protection to residential houses in Oakland, CA. This approach can be used in constructing an EP curve for any other community or region that faces potential losses from earthquakes or other disasters (e.g. Istanbul, Turkey). The losses could include residential structures, commercial and other buildings as well as infrastructure. As an alternative to viewing the EP curve through the eyes of an insurance company, one could take the position of the federal government who is concerned with the chances that an event will cause financial losses to the region greater than a certain amount.

[2] The last major published study undertaken by structural engineers to estimate damage ratios was ATC-13, published by the Applied Technology Council in 1985. In view of the extensive building damage experienced during the 1994 Northridge earthquake, the insurance industry would welcome a confirmation or update of that study by the structural engineering profession.

The Alpha Insurance Company has a book of business (BOB) which consists of a set of wood-frame homes in different parts of the city. Homes constructed prior to 1940 are assumed **not** to have crippled walls and **not** to be bolted to the foundations. The damage to these homes from an earthquake can be reduced if this risk mitigation measure (RMM) were adopted in the future. The distribution of structures for the Alpha Company is given below in Table 1. Structures whose age was unknown are assumed to fall into the Pre-1940 or Post-1940 era with the same likelihood as for the known structures.[3]

Table 1: Composition of Books of Business for Alpha Insurance Company

Don't Know	259
Pre-1940	3,091
Post-1940	1,650
Total	5,000

The EP curve for the Alpha Insurance Company with a 95% and 5% confidence intervals is depicted in Figure 2. For Oakland, we begin with a mean loss (the middle curve), the best estimate for all the parameters used in the loss estimation process. The mean loss curve is defined the most likely scenario of events.

The two parameters, frequency (F) and vulnerability (V), are varied in two ways relative to the mean loss, either high (H) or low (L). Based on the assumption that the two curves for F and V are on the high side and the two curves for F and V are on the low side with the other parameters at their base case, a 90% confidence interval using the joint distribution for F and V were generated. In other words, these estimates will cover the true estimate of the model parameter(s) with probability 0.90. The 95% confidence curve is a conservative estimate of more damage and the 5% confidence curve is an optimistic estimate of less damage.

Specifically, for the combination of parameters F and V, the 5% confidence curve is constructed by finding a pair of values ($F05$, $V05$), such that there is only a 5% chance that the true value of both parameters will be less than ($F05$, $V05$). Assuming that F and V are independently distributed, the required joint probability is:

[3] Thus, 169 of the 259 unknown structures in Company S's book of business (BOB) were assumed to be constructed prior to 1940 reflecting the ratio of pre-1940 to all known structures (3091/(1650+3091) in their BOB. These 169 were therefore eligible for mitigation.

$$P\{F < F05 \text{ and } V < V05\} = P\{F < F05\} \times P\{V < V05\} = 0.05$$

There are, of course, an infinite number of ways to pick *F*05 and *V*05 to make this equality true. We arbitrarily picked *F*05 and *V*05 so that each component had roughly the same marginal probability (i.e. *F*05=*V*05=.224)[4]. A similar analysis was used to construct the 95% confidence curve. (Grossi et al 1999)

Figure 2: EP Curves with 5% and 95% Confidence Intervals: Oakland, CA

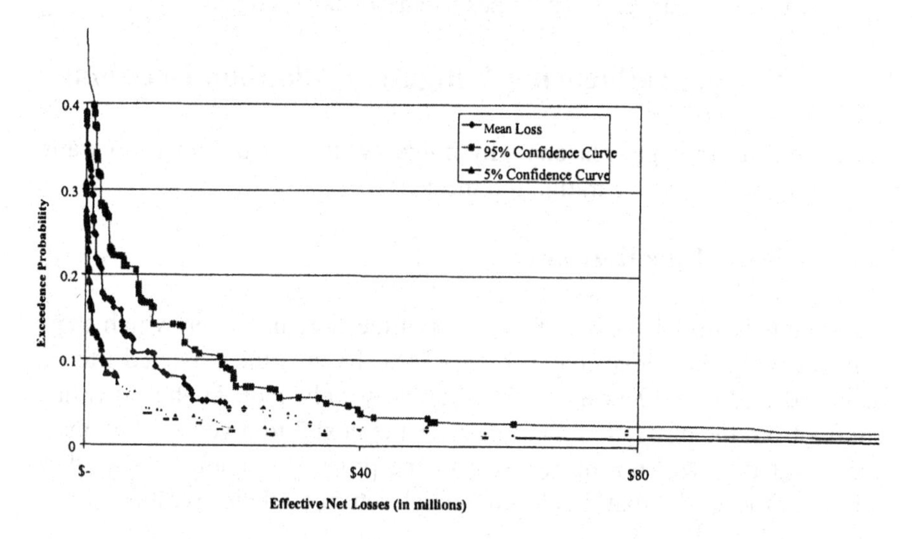

3. HOMEOWNER DECISION PROCESSES

In order to evaluate alternative risk management strategies for an insurance company, it is important to understand when homeowners are willing to invest in a cost-effective risk mitigation measure (RMM) voluntarily. The definition of cost-effective utilized here is related to the damage to the structure itself. In other words, we are interested in whether the expected discounted benefits with respect to loss reduction from the house exceed the cost of the measure itself. If a mitigation measure meets this criterion then it will certainly be viewed as desirable from a broader perspective when one takes into account other direct and indirect benefits, such as saving lives and the joys of not being forced to leave one's home.

There is a growing empirical literature, which provides insight into the typical individual in an earthquake-prone who has to determine whether or

[4] In other words F05 x V05= $(.224)^2$ =.05

not to invest in a protective measure. In a 1989 survey of 3,500 homeowners in four California counties subject to earthquake damage, only between 5 and 9 percent of the respondents in each of these counties reported adopting any RMMs (Palm et. al. 1990). A follow-up survey of residents affected by the October 1989 Loma Prieta earthquake and the Northridge earthquake of 1994 revealed that only 10 percent of homeowners invested in any type of structural loss-reduction measure whether or not they were affected by recent earthquakes in the State (Palm 1995). Measures such reinforcing crippled walls and bolting a house to a foundation may cost $1,000 to $3,000 but yield sufficient expected benefits to be justified for many existing homes in earthquake prone areas.

3.1 Factors Influencing Mitigation Adoption Decisions

There are five principal reasons why homeowners do not appear to want to invest in cost-effective mitigation measures:

3.1.1 Short Time Horizons

Individuals may have relatively short time horizons over which they want to recoup their investment in an RMM. Even if the expected life of the house is 25 or 30 years, the person may only look at the potential benefits from the mitigation measure over the next 3 to 5 years. They may reason that they will not be residing in the property for longer than this period of time and/or that they want a quick return on their investment.

3.1.2 High Discount Rates

The need for a quick return is also consistent with having a high discount rate regarding future payoffs. Loewenstein and Prelec (1992) propose a behavioral model of choice whereby the discount function is hyperbolic, rather than exponential. Their model appears to explain the reluctance of individuals to incur the high immediate cost of energy-efficient appliances in return for reduced electricity charges over time (Hausman, 1979; Kempton and Neiman, 1987).

3.1.3 Underestimation of Probability

Some individuals may perceive the probability of a disaster causing damage to their property as being sufficiently low that the investment in the protective measure will not be justified. For example they may relate their perceived probability of a disaster (p) to a threshold level (p*), which they may unconsciously set, below which they do not worry about the consequences at all. If they estimate $p < p^*$, then they assume that the event "will not happen to me" and take no protective actions.

3.1.4 Aversion to Upfront Costs

If people have budget constraints then they will be averse to investing in the upfront costs associated with protective measures simply because they feel they cannot afford these measures. It is not unusual for one to hear the phrase " We live from payday to payday" when asked why a household has not invested in protective measures. (Kunreuther et al. 1978).

3.1.5 Truncated Loss Distribution

Individuals may have little interest in investing in protective measures if they believe that they will be financially responsible for only a small portion of their losses should a disaster occur. If their assets are relatively limited in relation to their estimated potential loss, then these individuals may feel they that they can walk away from their destroyed home without being financially responsible. Similarly if residents anticipate liberal disaster relief from the government should they suffer damage, then they would have less reason to invest in a cost-effective mitigation measure.

3.2 Application to Oakland

Homeowners who are residing in pre-1940 structures in Oakland can consider bracing their crippled wall and bolting the house to the foundation. Mitigation costs are based on 1.2% times the cost of the structure. Thus for a $200,000 home in Berkeley, it would cost $2400 to brace the crippled wall and bolt the house to the foundation.

In this analysis of decision processes with respect to homeowners adoption of mitigation measures I focus on the impact of the first two factors---short time horizons and high discount rates and the homeowners' decision processes. Three scenarios will be considered to determine the cost-effectiveness of mitigation. Scenario 1 reflects a homeowner who has a long time horizon with low discount rate (20 years, 7%). Scenario 2 has a short time horizon and a low annual discount rate (5 years and 7%) while Scenario 3 has a short time horizon and high annual discount rate (5 years and 20%).

For each ZIP code in Oakland the cost effectiveness of these RMMs was determined by asking the following question: If every pre-1940 home were required by law to adopt the RMM, would the aggregate reduction in expected losses across the ZIP code be greater than the total annualized mitigation costs of all properties in the ZIP code which had been mitigated?

Using Scenario 1 all ZIP codes in Oakland would satisfy this cost effectiveness condition. By reducing the time horizon to 5 years, as in Scenario 2, then 68.8% of the ZIP codes in Oakland satisfied this cost effectiveness condition. For Scenario 3, where the discount rate is 20% and the time horizon is 5 years, only 37.5% of the ZIP codes in Oakland satisfied this cost effectiveness condition.

This analysis implies that if the decision to implement a building code in a zip code was based on an analysis of benefits and costs using short time horizons and/or high discount rates (Scenario 3), then the percentage of pre-1940 homes in Oakland that would have braced crippled walls and be bolted to its foundation would be only 37.5% of those that should have adopted this RMM.

4. INSURER DECISION PROCESSES

A literature has developed in recent years which suggests that insurers and other firms are risk averse and hence they must be concerned with non-diversifiable risks such as the possibility of catastrophic losses from disasters [Mayers and Smith (1982)]. Insurers are also likely to be ambiguity averse in that they are concerned with the uncertainty regarding the probability of a loss occurring [Kunreuther, Hogarth and Meszaros (1993)].

4.1 Importance of Safety First Constraint

Actuaries and underwriters both utilize heuristics that reflect these concerns. Actuaries normally determine a premium based on expected value by assuming that the probability and loss are known. They then increase this value to reflect the amount of perceived ambiguity in the probability and/or uncertainty in the loss. One commonly used formula for determining a premium is $z=(1+\lambda)\mu$ where $\mu=$ expected loss (i.e. $p \times L$) and $\lambda>0$ is a factor reflecting ambiguity and uncertainty independent of any adjustment to cover administrative costs [Lemaire (1986)].

Underwriters make their decision regarding whether a risk is insurable by utilizing the actuary's recommended premium z as a reference point and then focus on the impact of a major disaster on the probability of insolvency. In other words, underwriters are first concerned with the firm's safety and then with profit maximization. Stone (1973) formalized these concepts by suggesting that an underwriter who wants to determine the conditions for a specific risk to be insurable will first focus on keeping the probability of insolvency below some threshold level (q^*). More specifically, suppose that the insurer expects to sell m policies, each of which can create a loss L. Then the underwriter will recommend a premium z^* so that the probability of insolvency is no greater than q^*. Risks with more uncertain losses or greater ambiguity will cause underwriters to want to charge higher premiums for a given portfolio of risks. The situation will be most pronounced for highly correlated losses, such as earthquake policies sold in one region of California.

The empirical evidence based on surveys of underwriters supports the hypothesis that insurers will set higher premiums when faced with ambiguous probabilities and uncertain losses rather than a well-specified risk. In a survey of underwriters pricing the earthquake insurance, Kunreuther et al. (1995) showed that for the case where both the probability of a loss was ambiguous and the resulting loss uncertain, the premiums were between 1.43 to 1.77 times higher than if underwriters priced a non ambiguous risk.

4.2 Application to Oakland

Table 2 specifies the base case parameters for the Alpha insurance company providing coverage to homeowners in Oakland. We assume that full insurance coverage against damage from the disaster is available, with a 15 percent deductible. The (annual) premium charged is proportional to the expected loss (per year) for the property covered[5] and then multiplied by a loading factor (in this case 1.0) to reflect the administrative costs associated with marketing and claims settlement. In other words, for this analysis, property owners are charged premiums that are twice the expected losses to the insurer.

Table 2: Base Case Alpha Insurance Company Parameters

Parameter	Base Case Value
Company Assets	$ 4.5 Million
Deductible %: (expressed as a fraction of the value of property)	10%
Insurance loading factor:	1.0

Insurers are concerned with insolvency and focus on worst case scenarios in determining their portfolio of risks. A worst case loss (WCL)

[5] Expected loss to the insurer is defined as the probabilities of disasters of different magnitudes, each multiplied by the damage sustained minus the deductible and then summed.

is defined as a disaster where the probability of exceeding this dollar amount is some predetermined target ruin probability. For example, if an insurer sets a "target ruin probability" at .01, this implies that it would like to limit its book of business (BOB) so as to have at least a 99 percent chance of being able to pay insured losses from assets and premiums. If it has sufficient assets and/or premiums, then the insurer's actual probability of insolvency may be less than the this target level. Insurers who are more risk averse and hence have a greater concern with insolvency would reduce their target level probability to a lower level, say .002 (e.g. 1 in 500) which would imply a smaller BOB for a given asset base.

5. DEVELOPING A PLAN FOR DISASTER MANAGEMENT

There are several different strategies for reducing future losses from natural disasters and providing protection for victims of natural disasters that complement each other. This section briefly describes several policy tools for achieving these objectives.

5.1 Specific Policy Tools

We will be examining three policy tools as part of a disaster management strategy: building codes, indemnity contracts such as reinsurance and indexed catastrophe bonds. These options involve a number of different interested parties each of whom has their own values and objectives. The challenge is to develop a strategy for implementation which has enough positive returns to these different stakeholders for them to want to play the game.

5.1.1 Well-Enforced Building Codes

Building codes mandate that property owners adopt mitigation measures. Such codes may be desirable when property owners would otherwise **not** adopt cost-effective risk mitigation measures (RMMs) because they either misperceive the benefits from adopting the measure and/or underestimate the probability of a disaster occurring. For example, suppose the property owner believes that the losses from an earthquake to the structure is $20,000 and the developer knows that it is $25,000 because it is not well constructed. There is no incentive for the developer to relay the correct information to the property owner because the developer is **not** held liable should a quake cause damage to the building.

If the insurer is unaware of how well the building is constructed, then this information cannot be conveyed to the potential property owner through a premium based on risk.

There are many other players who are involved with implementing building codes. Banks and financial institutions could require an inspection of the property to see that it meets code before issuing a mortgage. Similarly insurers may want to limit coverage only to those structures that meet the building code. Either banks or insurers can provide a seal of approval to each structure that meets or exceeds building code standards. Inspecting the building to see that it meets code and then providing it with a seal of approval provides accurate information to the property owner on the condition of the house. It also signals to others that the structure is disaster resistant. This new information might then be translated into higher property values if prospective buyers took the earthquake risk into consideration when making their purchase decisions.

One way for communities to encourage well-enforced building codes is to provide tax incentives for more disaster-resistant homes. For example, if a homeowner reduces the chances of damage from an earthquake by installing a mitigation measure, then this taxpayer would get a rebate on state taxes to reflect the lower costs for disaster relief. Alternatively, property taxes could be reduced for the same reason. In practice, communities often create a monetary disincentive to invest in mitigation. A property owner who improves a home by making it safer is likely to have the property reassessed at a *higher* value and, hence, have to pay higher taxes. California has recognized this problem, and in 1990 voters passed Proposition 127, which exempts seismic rehabilitation improvements to buildings from reassessments that would increase property taxes.

The city of Berkeley has taken an additional step to encourage homebuyers to retrofit newly purchased homes by instituting a transfer tax rebate. The city has a 1.5 percent tax levied on property transfer transactions; up to one-third of this amount can be applied to seismic upgrades during the sale of property. Qualifying upgrades include foundation repairs or replacement, wall bracing in basements, shear wall installation, water heater anchoring, and securing of chimneys. Since 1993, these rebates have been applied to 6,300 houses, representing approximately $ 4.4 million in foregone revenues to the city (Earthquake Engineering Research Institute, 1998).

5.1.2 Indemnity Contracts

One way for private insurers to obtain protection against catastrophic losses is for them to purchase an indemnity contract against claim payments above a certain amount. A common indemnity contract is excess-of-loss reinsurance that provides coverage against unforeseen or extraordinary losses to the insurer. Specifically, the reinsurer charges a premium to indemnify the insurance company against all or part of the loss it may sustain under its policy or policies of insurance above a certain level.

For all but the largest primary insurers, a reinsurance-tied strategy is a prerequisite for offering insurance against hazards where there is the potential for catastrophic damage. An unusually severe set of claim payments can make even a well-capitalized insurer insolvent even if the insurer is, on average, profitable. A natural disaster with intense local effects, such as an earthquake, thus raises problems for insurers who cover multiple customers in a given geographic area because of the high correlation among the losses in their portfolio.

Reinsurers have similar concerns to those of the insurers and hence will limit their exposure in catastrophe-prone areas. A typical excess loss reinsurance contract requires the primary insurer to retain a specified level of risk and then covers all losses between an attachment point (L_A) and exhaustion point (L_E). In other words, the indemnity contract is of the following form: the reinsurer pays all losses in the interval L_A to L_E with a maximum payment of $L_E - L_A$.

If insurers were allowed to charge higher premiums on their own policies, many would need less reinsurance and would accept a higher attachment point L_A. However, regulatory constraints, such as obtaining prior approval by the State Insurance Commissioner on rate changes, limit insurers ability to raise premiums to levels that they feel reflect the risk. For example, in Florida following Hurricane Andrew in 1992 there were restrictions placed on rates that could be charged on homeowners coverage (which covers wind damage) in areas of the State affected by hurricanes (Lecomte and Gahagan 1998).

5.1.3 Catastrophe Bonds

As an alternative or complement to reinsurance, the insurer may want to utilize catastrophe-linked bonds (henceforth referred to as *cat bonds*) for protection. A cat bond requires the investor to put money up front, which could be used to pay for claims if some type of triggering event were to occur.

This process is called *securitization*, which simply means converting a financial contract into a security that can be traded in the secondary market. In practice, Alpha would begin the securitization process by meeting with investment banks who would provide Alpha with their estimates of the current market price of a cat bond. The investment bank would use risk assessment data from modeling firms to determine the expected losses from the specific cat bond that Alpha is interested in issuing. The expected losses along with spreads for comparable risks in the credit markets will determine the price of the cat bond. (Kuzak 2000)

In contrast to reinsurance where the reinsurer can become insolvent if it suffers catastrophic losses, the insurer does not face any credit risk from the cat bond. The money to pay for the losses is already in hand (usually deposited in escrow, and invested in short-term liquid securities). The first cat bond was issued by USAA in June 1997 to protect itself against cat losses from hurricanes in Florida. In this cat bond, USAA offered two tranches, geared towards different types of investors. Tranche one paid only a modest interest-rate premium above the risk-free rate (LIBOR), but investors would lose only their interest payments if USAA suffered hurricane losses during a 15-month period that exceeded $1 billion. Tranche 2 offered a higher premium over LIBOR but the investors' entire principal was at risk in case of severe hurricane losses by USAA.[6]

Most of the cat bonds which have been issued since the USAA offering are tied to a loss index (e.g., total insured losses from an earthquake in California) or to a disaster severity index (e.g., paying amounts for earthquake damages based on the Richter-scale measurements at specific locations in Japan) rather than to the insurer's losses.[7] If the index is independent of actual losses (as in the case of a disaster severity index), the insurer cannot manipulate the claims. Hence claims to insurers can be made immediately after the disaster rather than being subject to a time delay as in the case of reinsurance.

On the other hand, such a cat bond may create **basis risk**. Basis risk refers to an imperfect correlation between the actual losses caused to the insurer and the payments received from the cat bond. Traditional excess-of-loss reinsurance has zero basis risk because there is a direct relationship between the loss and the payment delivered by the reinsurance instrument. Even a cat bond based on some verifiable, non-manipulable index (e.g., aggregate insurance industry losses, the Richter or Saffir-Simpson scales, total rainfall in Rangoon during August) is subject to basis risk. In other words, the insurer's book of business may not be accurately represented by the index, and therefore the insurer's losses will not be perfectly correlated with the actual payments from the cat bond triggered by the index.[8]

---

[6] For more detail about USAA's financing decision, see Froot and Seasholes (1997)..

[7] For more details on the structure of recent cat bonds see Insurance Services Office (1999).

[8] See Major (1999) for a more detailed description of basis risk, and the effect of basis risk on insurers' ability to address catastrophic losses.

6. A RISK MANAGEMENT STRATEGY FOR OAKLAND, CALIFORNIA

The model city of Oakland, CA offers an opportunity to examine the impact of alternative disaster management strategies on the performance of the insurer. More specifically we will be examining the role of building codes, reinsurance and indexed catastrophe bonds on the performance of the Alpha Insurance Company.

6.1 Specific Policy Tools

The following policy tools will be examined: Building codes, Indemnity contracts and Catastrophe Bonds.

6.1.1 Building codes

We will compare the losses to Alpha for two polar cases: (1) when there are no building codes in place so that all pre-1940 homes are not mitigated and (2) all homes have crippled walls and their foundations braced.

6.1.2 Indemnity Contracts

Alpha will purchase an excess of loss reinsurance contract that has an attachment point (L_A) where there is a 4 percent chance that the losses will exceeding this amount when there is no mitigation in place. The exhaustion point (L_E) is determined so that the chances of Alpha's losses exceeding this amount is 3 percent when no pre-1940 homes are mitigated. More specifically Alpha's reinsurance contract has a value of L_A = \$5.93 million and L_E = \$8.27 million. Alpha must pay a premium that reflects the risks that the reinsurer faces plus a loading factor of 100 percent. Thus if the expected losses to the reinsurer over the interval L_E - L_A was \$40,000, then Alpha would pay the reinsurer \$80,000 in premiums.

6.1.3 Catastrophe Bonds

Insurers can issue a parameterized indexed cat bond where they would receive payments according to a predetermined measurement schedule. More specifically Alpha issues a catastrophe bond that pays investors an interest premium in exchange for guaranteed funds based on the occurrence of a disaster of a given intensity or magnitude. The amount of funds given to the insurer is based on an index (e.g. an earthquake of 7.5 on the Moment Magnitude scale) so it likely will not be perfectly correlated to actual claim payments.

Suppose that Alpha issues a \$10 million Total Face Value cat with a 10% annual coupon. If the risk-free interest rate (LIBOR) is 5.5% then the spread above this rate for this cat bond is 4.5% (10%-5.5%). At the beginning of the year Alpha will receive \$10 million from investors and immediately reinvest this amount in a risk free investment earning \$550,000 (i.e., 5.5% x \$10 million). Insurers will pay investors guaranteed coupon payments of \$1 million (i.e., 10% x \$10million) so that the price to the insurer for interest rates over LIBOR is \$450,000 (\$ 1 million-\$550,000)

In return for a higher return the investor does face the possibility of losing some or all of its principal if a severe enough earthquake occurs in Oakland. The amount paid out to the insurer depends on how the cat bond is constructed. The cat bond was designed to protect insurer from losses greater than the reinsurance exhaustion point (L_E). We constructed the cat bond by first computing the expected losses generated by earthquakes of different intensities in Oakland. For example, if one specified all earthquakes of magnitude 7.0 to 7.5 that would impact residential homes in Oakland, CA we could determine the probability of each of these events i ($p_{i\ (7.0\text{-}7.5)}$) occurring and the damage that each event ($L_{i\ (7.0\text{-}7.5)}$) would create. The expected loss for earthquakes of this magnitude would be

$$E\{L_{i(7.0-7.5)}\} = \frac{\sum_i p_{i(7.0-7.5)} L_{i(7.0-7.5)}}{\sum_i p_{i(7.0-7.5)}}$$

In the above equation, we only used losses in excess of L_E since we wanted to model the cat bond to protect the insurer from losses in the far right tail of the distribution. We did this same type of calculation for each earthquake in Oakland within specified intervals (e.g. 7.0-7.5 MMI). We then set these expected losses equal to the actual payout of the cat bond to the insurer for acceptable ranges of magnitudes. More specifically, cat bonds paid the insurer for earthquakes greater than 7.0 in Oakland. These payouts for cat bonds will create basis risk since there will be some events in the interval that will produce insurer claims (net of reinsurance payments) larger than the amount provided by the cat bonds (positive basis risk) and those that are smaller than the cat bond amount (negative basis risk).

Table 3 depicts the structure of two different cat bonds for Oakland depending on whether or not a building code was in place. When there were no pre-1940 homes mitigated, then the total face value of the cat bond was $30 million and the price to the insurer for a higher interest rate than LIBOR was $1.342 million. When homes were mitigated then the insurer only required a cat bond with face value of $18.5 million, since catastrophic losses were reduced. As shown in the payout schedule for the cat bonds, when there was no mitigation in place insurers would receive $29.196 million if an earthquake occurred in Oakland that had a magnitude between 7.0 and 7.5 on the MMI scale. Interestingly enough the payout was only $1.058 million when earthquakes greater than 7.5 registered on the scale. The smaller figure was due to the relatively few residential structures in those portions of Oakland where earthquakes of this magnitude or greater could occur.

Table 3: Structure of Cat Bond: Oakland

(All Dollar Amounts in $ Thousands)

Parameters:	0% Mitigation	100% Mitigation
Total Face Value	$30,000	$18,500
Cat Bond Return to Investor	9.97%	10.00%
Price to Insurer over LIBOR	$1,342	$832

		0% Mitigation	100% Mitigation
Magnitude	*Probability*	*Payout to Insurer*	*Payout to Insurer*
< 7.0	98.0%	$0	$0
[7.0,7.5)	1.8%	$29,196	$18,279
[7.5,max)	0.2%	$1,058	$203

6.2 Performance of Alternative Risk Management Strategies

There are eight different strategies which could be undertaken by Alpha using the above three policy tools depending on whether or not one had well-enforced building codes, reinsurance and cat bonds in place.[9] Building codes and cat bonds will lower the exceedance probability curves for Alpha from the strategy of no building code and no available cat bonds. Building codes reduce the damage to pre-1940 homes from earthquakes. Cat bonds provide insurers with pre-determined payments as a function of the magnitude of the earthquake and independent of the actual damage. Hence they will reduce the losses over what they otherwise might have been. Insurers have to pay for this protection in the form of interest rates substantially above LIBOR. Figure 3a depicts the EP curves with and without cat bonds when **no** mitigation is in place for pre-1940 homes. Figure 3b examines the EP curves with and without cat bonds when a building code is in place for pre-1940 homes.

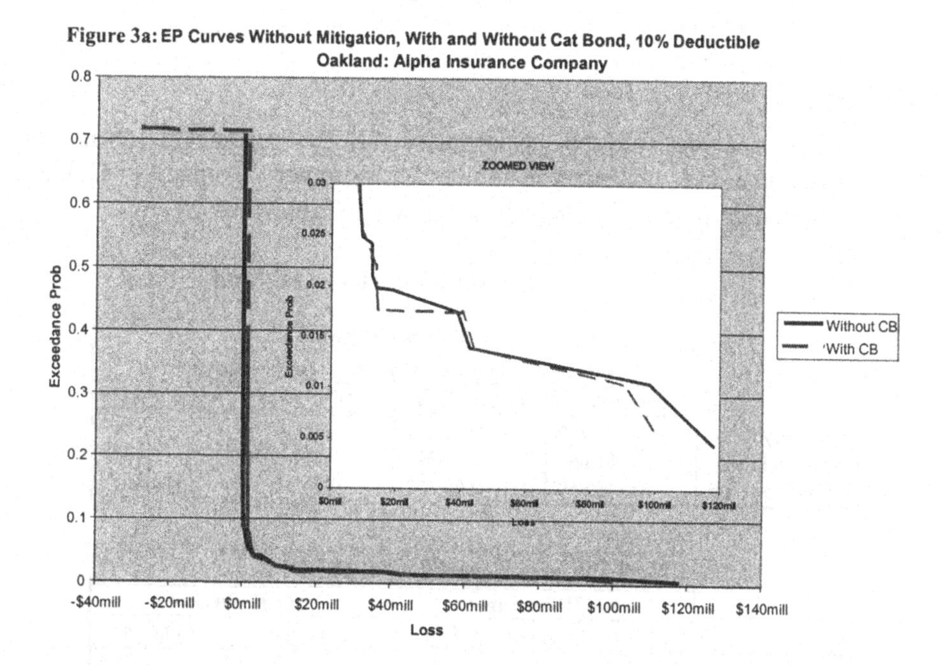

[9] These eight strategies reflect the eight possible combinations of the 3 different policy tools (building codes/no building codes; reinsurance/no reinsurance; cat bonds/ no cat bonds). For example one strategy would be "Building codes, Reinsurance, no Cat Bonds".

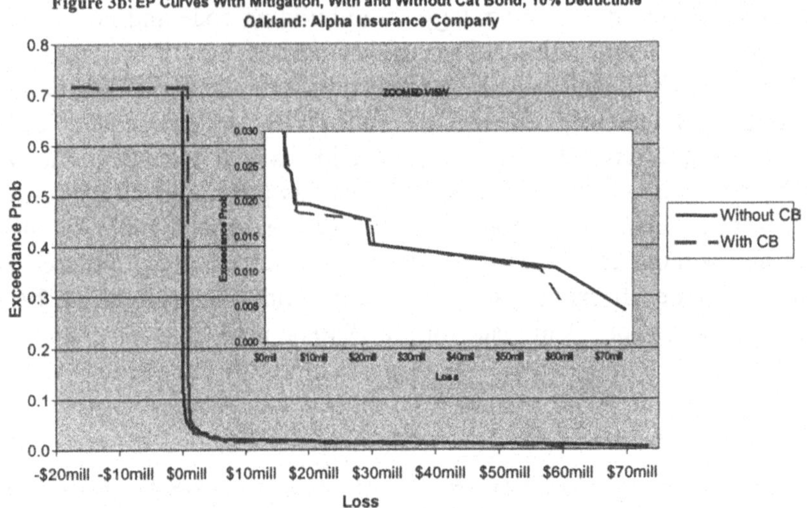

The analyses of the performance of the impact of mitigation, reinsurance and cat bonds on the Alpha Insurance Company in Oakland reveals some interesting findings as shown in Table 4.

Table 4: Summary Results - Oakland

	0% Mitigation			
	w/o Reinsurance		w/ Reinsurance	
	w/o Cat Bond	*w/ Cat Bond*	*w/o Cat Bond*	*w/ Cat Bond*
Insolvency Prob	3.19%	2.87%	2.44%	2.22%
Expected Profit	$1,680	$898	$1,604	$822
Worst Case Loss	($114,660)	($96,173)	($112,468)	($93,981)

	100% Mitigation			
	w/o Reinsurance		w/ Reinsurance	
	w/o Cat Bond	*w/ Cat Bond*	*w/o Cat Bond*	*w/ Cat Bond*
Insolvency Prob	1.98%	1.89%	1.98%	1.76%
Expected Profit	$954	$473	$905	$424
Worst Case Loss	($71,449)	($58,254)	($69,990)	($56,794)

6.2.1 Impact of Mitigation

An analysis of the expected loss reduction for Oakland reveals that bracing the crippled wall and bolting pre-1940 houses to their foundation is a cost-effective RMM. For very severe earthquakes in Oakland this mitigation measure reduces the worst case losses (WCL) by approximately 40% whether Alpha purchases reinsurance and/or cat bonds.

6.2.2 Impact of Reinsurance

As expected, reinsurance reduces the insolvency probability of Alpha but its expected profits are also reduced because Alpha is assumed to have 100 percent BOB. If the insurer were able to write more coverage because it had reinsurance, then profits could actually be higher. In other words, the reinsurance would expand the insurer's capacity.

6.2.3 Impact of Cat Bonds

Turning to the impact of cat bonds on the insolvency and profitability of Alpha, we obtain the following results, Cat bonds reduce the insolvency probability by almost 0.2 percent when there is no reinsurance or mitigation in place and reduced the insolvency probability by 0.1 percent when both reinsurance is purchased and mitigation is in place. Cat bonds reduced WCLs by about $20 million with no mitigation and by about $10 million with mitigation.

6.3 Conclusions

In summary, a combination of building codes, reinsurance and indexed cat bonds can form a useful strategy for reducing losses to property owners as well as insurers and the investment community. The implementation of this strategy requires a concerted effort by both the public and private sectors. For example, building codes require inspections by certified personnel. Banks and financial institutions can help enforce these building codes by making loans conditional on such an audit. Insurers can offer lower premiums for those adopting cost-effective mitigation measures.

With new sources of capital from the Bermuda market, there is an opportunity for reinsurance to provide more protection against insurers' potential losses if the premiums are attractive enough to them. With respect to new financial instruments, the interest rate on cat bonds has to be sufficiently low so that insurers will want to issue them but high enough for investors to want to purchase them.[10] One needs to determine

---

[10] See Bantwal and Kunreuther (2000) for some of the reasons why cat bonds are priced as high as they are today.

what the appropriate role of the public sector is in providing financial protection against large losses. The reluctance of the insurance industry to cover losses from earthquakes in California led to the formation of the California Earthquake Authority which is a state-run insurance company funded by the insurance and reinsurance industry will limited liability.

The potential for developing an effective disaster management strategy has been made possible by the new advances in information technology (IT) and risk assessment over the past ten years. More sophisticated catastrophe models have provided far more accurate estimates of the likelihood and potential losses from future natural disasters than we have had in the past. These models coupled with user-friendly software have paved the way for an analysis of the cost-effective of mitigation measures as well as the emergence of new capital market instruments. Our own work on the Wharton Catastrophic Risk project would not have been possible without the aid of these models.

7. PROPOSED FUTURE WORK: A STUDY OF ISTANBUL, TURKEY

There is much to be done in developing strategies for reducing losses and providing financial protection against natural disasters. The specific programs will depend on the institutional arrangements for specific countries. A group of researchers from Turkey and the United States are planning to engage in a study of natural hazards management by focusing on the risks of earthquakes in Istanbul, Turkey.

7.1 Research Objectives

The project will examine the nature of the uncertainties surrounding risk estimates and the needed data and infrastructure to support this analysis. This research should enable us to gain further insight into how mitigation coupled with financial instruments can reduce losses and provide funds for recovery from a mega-disaster in an emerging economy. Although the focus is on the earthquake hazard, the concepts will be applicable to other natural hazards that have catastrophic potential, such as floods and hurricanes.

The research will address the following three questions:

(1) How can one utilize risk assessment methodologies for estimating the potential damage and the uncertainties surrounding these figures for earthquakes of different magnitudes and intensities?

(2) What role can mitigation measures and new financial instruments play in reducing losses from future disasters and providing funds for recovery?

(3) How can one utilize model cities for evaluating the linkages between risk managements strategies such as investments in risk reduction (mitigation) and risk transfer (new financial instruments) for dealing with large-scale disasters?

Istanbul, Turkey has been selected as a model city since it is representative of many metropolitan areas in emerging economies and has the potential for severe losses should a major earthquake occur in the future. The Kocaeli earthquake in August 1999 caused over 19,000 confirmed fatalities with a total economic loss estimated to be at least $15 billion (Hurriyet, 1999). This event led to a recognition by Turkey's government, industry and the public for the urgent need to develop and enforce better building standards, to promote improved business contingency planning for supply distortions and to provide more effective emergency preparedness and response. There is a recognized need for developing new financial instruments to aid the recovery effort given the limited role that insurance has played up-to-date in post-disaster funding. For example, after the recent Kocaeli earthquake, out of the 500,000 housing units that were adversely affected, only 1000 were privately insured (METU-DMC, 1999).

7.2 Research Plan

The proposed research will be guided by a framework which links the process of characterizing the risk from natural disasters with a set of private and public sector initiatives and strategies for reducing future losses and providing funds for recovery. It will involve the following tasks:

7.2.1 Construct Model City and Undertake Risk Assessment

We will first structure the data to characterize the nature and potential consequences of seismic hazards for Istanbul composed of residential structures, industrial and commercial facilities as well as infrastructure (e.g. water, electricity lifelines, transportation). As part of the risk assessment process, we will incorporate uncertainty bands around damage estimates for different structures as well as for total losses using the standard Exceedance Probability (EP) curves to reflect ranges of losses. This analysis enables us to determine the impact that uncertainty has on the role of mitigation, insurance and new financial instruments in managing risk. (Grossi et al. 1999).

7.2.2 Survey of Homeowner Attitudes toward Mitigation and Building Codes

This task involves undertaking a survey of homeowners in Istanbul that to determine on the role mitigation can play in dealing with the earthquake risk. It will focus on knowledge of and attitudes towards factors that may affect the adoption of risk mitigation measures against a possible earthquake. We also plan to undertake a series of interviews with building construction officials and engineers as well as municipal and government officials to determine their attitudes towards building codes and construction practices.

We will also determine what mitigation measures they are currently considering and their perceived cost-effectiveness for new and existing structures. In addition to the survey work, a set of experiments will be conducted in Turkey to study the individual decision processes concerning risk mitigation investments. These experiments will build on previous experimental work we have undertaken on individuals' decision processes regarding the adoption of protective measures.

7.2.3 Risk Management Tools for Reducing Risks from Natural Disasters

This task will evaluate the relevant roles of insurance and new financial instruments in providing protection against the earthquake risk. Istanbul will be used as a laboratory for this analysis.

We will first be examining the most efficient combination of financing instruments to provide funds for recovery, as well as to encourage individuals to take protective actions prior to a disaster. In the United States insurance is an option that many individuals can take advantage of but often do not utilize. Similarly few individuals invest in mitigation measures voluntarily. (Kunreuther and Roth, 1998). In emerging economies such as Turkey, there is limited insurance available but extended family and tight community structures provide informal insurance markets (Arnott and Stiglitz, 1990). Mitigation has not been an issue until after the recent Kocaeli earthquake when so many buildings collapsed in part due to poor building code enforcement.

We will then turn to the types of funding that can be provided to residents in hazard-prone areas affected by a major disaster, focusing primarily on insurance and capital market financing. We will then examine the increasing opportunities for using the capital market to facilitate insurance and/or to substitute for it and the role that mitigation can play in reducing the price of these instruments and the need for them. The capital market can provide capital or capital substitutes such as insurance- linked securities that enhance the capacity of the insurance market. The capital market can also provide hedge or capital enhancement products directly to Third World government and corporations or directly to development banks and related institutions (Croson and Richter 1999).

7.3 Lessons from Istanbul

This research should enable one to examine the impact of different risk management tools for mitigation and recovery as a function of the institutional arrangements within the city and country. In particular, we will be comparing the impacts of earthquakes of different magnitudes on Istanbul under varying assumptions concerning mitigation measures adopted and financial instruments. Here are some of the questions we will want to address:

- How will the efforts of mitigation perform in Istanbul under earthquakes of different magnitudes? Are there lessons with respect to mitigation measures and building code enforcement in Istanbul that are relevant to the US?
- What is the performance of different financial instruments given earthquakes of different magnitudes? Given that Istanbul does not have a well-developed insurance industry it will have to rely primarily on financial instruments and governmental risk-bearing and disaster relief for dealing with future losses in the short run. What are the lessons from Turkey that may be transferable to other emerging economies in financing recovery from catastrophic losses?
- What can we learn from this exercise that could be helpful to US funding agencies, relief agencies and other governmental and non-governmental organization that may have an interest in protecting the viability of emerging economies as a market to the US economy?

We look forward to working with others on these important issues and benefiting from your insights as to how we can better manage losses from natural disasters both in the United States and emerging economies.

REFERENCES

Arnott, R., and J. Stiglitz. 1990. "The Welfare Economics of Moral Hazard." In *Risk, Information and Insurance: Essays in the Memory of Karl H. Borch*, edited by H. Louberge, Boston, MA: Kluwer Academic Publishers.

Bantwal, Vivek, and Howard Kunreuther. 2000 "A Cat Bond Premium Puzzle?" *Journal of Psychology and Financial Markets* 1:76-91.

Croson, D., and A. Richter. 1999. "Sovereign Cat Bonds and Infrastructure Project Financing." Working Paper 99-05-25, Wharton Risk Management and Decision Processes Center, University of Pennsylvania. Philadelphia, PA: Forthcoming, *Risk Analysis.*

Earthquake Engineering Research Institute (EERI). 1998. *Incentives and Impediments Improving the Seismic Performance of Buildings.* Oakland, CA: Earthquake Engineering Research Institute

Froot, K., and M. Seasholes. 1997. "USAA: Catastrophic Risk Financing." Case 9-298-007; Harvard Graduate School of Business Administration; Boston, MA.

Grossi P., P. R. Kleindorfer and H. Kunreuther. 1999. "The Impact of Uncertainty in Managing Seismic Risk: The Case of Earthquake Frequency and Structural Vulnerability", Working Paper 99-03-26, Risk Management and Decision Processes Center, University of Pennsylvania. Philadelphia, PA.

Hanks, Thomas, and C. Allin Cornell. 1994. Probabilistic seismic hazard analysis: A beginner's guide. *Proceedings of the Fifth Symposium on Current Issues Related to Nuclear Power Plant Structures, Equipment and Piping.* Raleigh: North Carolina State University.

Hausman, Jerry. 1979. " Individual Discount Rates and the Purchase and Utilization of Energy-Using Durables." *Bell Journal of Economics* 10, pp.33-54.

Heinz Center for Science, Economics, and the Environment. 1999. *The Hidden Costs of Coastal Hazards: Implications for Risk Assessment and Mitigation.* Washington, DC: Island Press.

Hurriyet, October 21st, 1999, p. 14.

Insurance Services Office 1999. *Financing Catastrophe Risk: Capital Market Solutions*, New York, N.Y.: Insurance Services Office

Kempton, Willett, and Max Neiman (Eds). 1987. *Energy Efficiency: Perspectives on Individual Behavior.* Washington, D.C.: American Council for an Energy Efficient Economy.

King, Stephanie, and Anne Kiremidjian. In press. "Use of GIS for earthquake hazard and loss estimation." In *Geographic Information Research: Bridging the Atlantic.* London: Taylor & Francis.

Kleindorfer, P. R., and H. Kunreuther. 1999. "Challenges Facing the Insurance Industry in Managing Catastrophic Risks." In *The Financing of Property/Casualty Risks*, edited by Kenneth Froot. Chicago: University of Chicago Press.

Kunreuther, Howard, et al. 1978. *Disaster Insurance Protection: Public Policy Lessons.* New York: John Wiley and Sons.

Kunreuther, Howard, Robin Hogarth, and Jacqueline Meszaros. 1993. "Insurer Ambiguity and Market Failure." *Journal of Risk and Uncertainty* 7: 71-87.

Kunreuther, Howard, Jacqueline Meszaros, Robin Hogarth, and Mark Spranca. 1995. Ambiguity and underwriter decision processes. *Journal of Economic Behavior and Organization* 26:337—352.

Kunreuther, Howard, and Roth, Richard, Sr. (Eds). 1998. *Paying the Price: The Status and Role of Insurance Against Natural Disasters in the United States*. Washington, D.C: Joseph Henry Press.

Kuzak, Dennis. 2000 "The Capital Markets." In *Financial Management of Earthquake Risk*, Earthquake Engineering Research Institute. Oakland, CA: EERI.

Lecomte, Eugene, and Karen Gahagan. 1998. "Hurricane Insurance Protection in Florida." In *Paying the Price: The Status and Role of Insurance Against Natural Disasters in the United States* edited by Howard Kunreuther and Richard Roth, Sr. Washington, D.C: Joseph Henry Press.

Lemaire, Jean. 1986. *Theorie Mathematique des Assurances*. Belgium: Presses Universitaires de Bruxelles.

Loewenstein George, and Drazen Prelec. 1992. "Anomalies in Intertemporal Choice" *Quarterly Journal of Economics*, 107, pp. 573-597.

Major, John A. 1999. "Index Hedge Performance: Insurer Market Penetration and Basis Risk." In *The Financing of Property/Casualty Risks*, edited by Kenneth Froot. Chicago: University of Chicago Press.

Mayers, David, and Clifford Smith. 1982. "On corporate demand for insurance." *Journal of Business* 55:281—296.

METU-DMC. 1999. "An Analysis of the Earthquake Vulnerability of the Housing Stock in Turkey." Middle East Technical University, DMC Report, Ankara.

Palm, Risa, Michael Hodgson, R. Denise Blanchard, and Donald Lyons. 1990. *Earthquake Insurance in California: Environmental Policy and Individual Decision Making*. Boulder, Colo.: Westview Press.

Palm, Risa. 1995. *Earthquake Insurance: A Longitudinal Study of California Homeowners*. Boulder, Colo.: Westview Press.

Stone, James. 1973. "A theory of capacity and the insurance of catastrophe risks: Part I and Part II." *Journal of Risk and Insurance* 40:231—243 (Part I) and 40:339—355 (Part II).

Wilczynsky, Piotr, and Vijay Kalavakonda. 2000. "TURKEY: Strategy for Disaster Management and Mitigation." Paper presented at World Bank ProVention Consortium Seminar. Washington, DC: February 2-4.

6

A Behavioral Perspective on Risk Mitigation Investments*

Ayse Öncüler
Decision Sciences, INSEAD

Decisions concerning risk mitigation generally involve tradeoffs between immediate versus delayed outcomes and certain versus risky ones. The decision in adopting a risk mitigation measure illustrates an intertemporal choice problem under uncertainty. This study explores this relationship between delay and uncertainty in risky intertemporal decision-making. In particular, we focus on the following questions: (1) Do risk preferences depend on the time period over which the outcomes are evaluated? (2) Does the discount factor (rate of time preference) depend on the riskiness of the outcomes being evaluated? (3) Is there an interaction between the effect of delay and uncertainty and, if so, in what direction is this interaction? Contrary to normative predictions, we find that delay and risk discounting are *not* independent from each other. In particular, the delay discount rate increases for uncertain future outcomes and the risk discount rate decreases with an increase in delay. Thus individuals are more impatient for gambles than for certain outcomes and less risk-averse (more risk-neutral) for delayed outcomes than for immediate ones. The findings also suggest that the simultaneous presence of delay and uncertainty leads to a higher discounting of risky future outcomes than predicted by either effect separately.

1. INTRODUCTION

Decisions concerning risk mitigation generally involve tradeoffs between immediate versus delayed outcomes and certain versus risky ones. For instance, consider a homeowner who has the option of bracing

* I'd like to thank Jonathan Baron, Rachel Croson, Jack Hershey, Paul Kleindorfer and Howard Kunreuther for their helpful comments and suggestions.

P.R. Kleindorfer and M.R. Sertel (eds.), Mitigation and Financing of Seismic Risks, 101–127.

the house to its foundation to reduce the chances of damage in case of an earthquake. This protective measure has an upfront cost and in return, it will decrease the probability of earthquake damage in the future. In this case, the homeowner has to decide whether to invest in this measure or not based on the trade-off between the setup cost and the probabilistic benefits over time.

The decision in adopting a risk mitigation measure against a possible earthquake illustrates an intertemporal choice problem under uncertainty. The decision-maker has to evaluate a future outcome that is risky in nature. As a result, the risk attitude of the decision-maker, in combination with his or her evaluation of the future, plays a role in reaching a decision. In such cases, it is crucial to study the interaction of delay (time preference) and uncertainty (risk preference) since these two effects may have an impact on the final outcome that is different than the impact of each effect separately. For instance, in the earthquake example given above, if the homeowner becomes less risk-averse as the length of time (s)he is planning to live in the house increases, he or she might be less willing to invest in the protective measure. Such an interaction between time and risk preferences is the main focus of this study.

In previous literature, the modelling and the elicitation of time and risk preferences has evolved as two separate research areas. Most of the existing literature focuses on either the impact of delay on risk attitudes or the effect of uncertainty on time preferences but not on the combination of the two. Even though the similarity between time and risk preferences (in terms of modelling and behavioural anomalies) has been reported in some studies (see e.g. Rachlin *et al.* 1986, Loewenstein and Prelec 1991, Rachlin and Siegel 1994), previous research on the interaction between delay and uncertainty effects has been quite limited. This study examines the effect of a *simultaneous* presence of delay and uncertainty on valuing future risky outcomes. In a set of controlled experiments, we examine the existence and direction of the interaction of risk and time preferences. In particular, we focus on the following research questions:

1. Do risk preferences depend on when the uncertainty is resolved? In other words, are risk preferences stationary or do they change as the resolution of a risky event is postponed to the future? If they do, in which direction is this change?
2. Does the discount factor (rate of time preference) depend on the riskiness of the outcomes being evaluated? If it does, in which direction does this rate change as the outcome becomes more risky?

3. Is there an interaction between delay and uncertainty effects? If so, what is the direction of this interaction and how does it influence the evaluation of risky future outcomes?

The paper proceeds as follows: In Section 2, the experimental model of risky intertemporal choice is introduced and the predictions are discussed. After describing the experimental design in Section 3, the findings are presented in Section 4. The paper concludes with a discussion of implications and possible directions for future research.

2. EXPERIMENTAL MODEL

2.1 Two Models of Risky Intertemporal Choice

This study focuses on the class of decision problems where an irrevocable commitment is made at $t=0$ and uncertain payoffs are distributed over time. The probability distribution of the future payoffs are fixed given the information at $t=0$. In the earthquake mitigation case, this would mean a homeowner decides to strengthen the house foundation in the current period and this determines the future probability distribution of damage. In order to test the separability of delay and uncertainty effects within this framework, two models of risky intertemporal choice, which are based on a natural extension of the existing models of decision-making under uncertainty and intertemporal choice, will be examined. The first extension is a combination of two normative models: discounted utility theory (Koopmans 1960; Samuelson 1982) and expected utility theory (von Neumann and Morgenstern 1944). Hereafter, we refer to this normative model as the *discounted expected utility model* (DEU).[1] The valuation functional of DEU model is given by the following formulation:

$$U(x) = \sum_t \sum_i \delta^t \cdot p_{it} \cdot u(x_{it})$$

where the future is discounted at an exponential rate δ and the utility of each outcome x_{it} is weighted according to the corresponding probability p_{it}.[2]

[1] See e.g. Lucas (1978) and Mehra and Prescott (1985) for different analytical applications of the DEU model.

[2] The relation between the discount factor δ and the discount rate r is given by $\delta^t = 1/(1+r)^t$.

Turning to our example of earthquake mitigation, in a DEU framework, a homeowner would be willing to pay for the protective measure up to *U(x)*, where

$$U(x) = \sum_t \sum_i \delta^t \cdot (p_{it} - p_{it}^*) \cdot u(x_{it})$$

where the difference between p_i and p_i^* is the reduction in the probability of a loss in case an earthquake happens, x_{it} is the loss in period t and δ is the discount factor.

Empirical studies have shown that individuals generally discount the future in a non-exponential fashion- slower in the initial periods and then faster (e.g. Thaler 1981, Benzion *et* al. 1989, Ainslie 1991). DEU fails to explain this non-stationary discounting. It also suffers from some other behavioural inconsistencies such as the violation of independence axiom (e.g. Allais and Ellsberg paradoxes). Taking into account such possible problems, a more behavioural framework could be proposed, which brings together the intertemporal model of Loewenstein and Prelec (1992) and the model of prospect theory (Kahneman and Tversky 1979). We refer to this behavioural model as *prospect theory/ intertemporal choice model* (PT/IC), which is formulated by

$$V(x) = \sum_t \sum_i \phi(t) \cdot \pi(p_{it}) \cdot v(x_{it})$$

where $\phi(t)$ is the hyperbolic discount function, $\pi(p_i)$ is the probability weighting function, $v(x_{it})$ is the value function specified in the prospect theory and X is the vector $(x_{11},...,x_{nT};\ p_{11},...,p_{nT})$. This model is built on the assumptions of prospect theory and the Loewenstein-Prelec theory of intertemporal choice. We should note that this model is only relevant for binary lotteries, which is the main focus in the study.[3]

In the earthquake protection case, an individual who maximizes PT/IC utility would pay up to *V(x)*:

$$V(x) = \sum_t \sum_i \phi(t) \cdot (\pi(p_{it}) - \pi(p_{it}^*)) \cdot v(x_{it})$$

3 Note that cumulative prospect theory (Tversky and Kahneman 1992) can handle more complicated lotteries.

where the difference between p_{it} and p_{it}^* is again the reduction in the probability of a loss in case an earthquake happens, x_{it} is the loss in period t and $\phi(t)$ is the discount function at period t.

Both DEU and PT/IC models assume that the effects of delay and uncertainty do not depend on each other. This follows from the fact that the discount rate and probability functions are separable in both of the models.[4] Since the effects of delay and uncertainty are separable, these models predict that the simultaneous effect of delay and uncertainty can be described by combining each effect separately. In this study, a set of experiments is conducted to test this assumption of separability that exists in both the normative and descriptive extensions of risky intertemporal decision-making.

Using DEU and PT/IC as the two benchmark models underlying the experiments, we derive two quantities of interest that will be the basis of the experiments: α (the delay discount rate) and β (the risk discount rate). In order to derive the delay and uncertainty discount rates, we first define the present value of a stream of risky prospects over time. First, we can assume that there is a time-induced vector of outcome random variables $x = (x_0, \ldots x_T)$, reflecting payoffs in periods $t=0$ (now) to $t=T$. When $x_0=0$, we will omit it for simplicity and refer to $x=(x_1,\ldots,x_T)$ as the vector.[5] In this case, the present value of x at $t=0$, $PV_0(x)$, can be defined as the certainty equivalent at $t=0$ of $x=(x_1,\ldots,x_T)$. In the DEU model, assuming $U(0)=0$ for convenience, $PV_0(x)$ would be characterized by

$$U(PV_0(x)) + \sum_{t=1}^{T} \delta^t EU(x_t) = 0$$

In the PT/IC model, again assuming $V(0)=0$ for convenience where 0 is also the reference point, $PV_0(x)$ would be characterized by

$$V(PV_0(x)) + \sum_{t=1}^{T} \phi(t) E_\pi V(x_t) = 0$$

where $E_\pi V(x_t)$ is given by

---

4 p_i and δ^t in *DEU* and $\pi(p_i)$ and $\phi(t)$ in *PT/IC* are separable.

5 In the earthquake measure, we assume at $t=0$ the decision-maker makes the investment and receives no payoff yet.

$$E_{\pi}V(x_t) = \sum_{i=1}^{n} \pi(p_{it})v(x_{it})$$

It will be useful to express $PV_0(x)$ in a manner which separates the effects of time and risk, focusing on lotteries with outcomes in a single future period, i.e. on x vectors of the form $x=(0,...,0, x_t, 0,...,0)$ with at most a single non-zero entry. For this case, we could denote $PV_0(x)$ as $PV_0(x_t)$, obtaining the following characterizations from the above formulations of DEU and PT/IC respectively:

$$U(PV_0(x_t)) = -\delta^t EU(x_t)$$

$$V(PV_0(x_t)) = -\phi(t)E_{\pi}V(x_t)$$

From these, we can now define two fundamental quantities of interest in the experiments that follows:

$\alpha(t,p)$ = delay discount rate when x_t is distributed according to the probability distribution p

$\beta(t,p)$ = risk discount rate when x_t is distributed according to the probability distribution p

The delay discount rate α shows the rate of discounting when an outcome is postponed to the future. Fixing some outcome random variable at time period t and assuming $PV_0(x_t) \neq 0$, we define α as follows:[6]

$$\alpha(t,p) = \frac{CE(x_t) - PV_0(x_t)}{PV_0(x_t)} \tag{1}$$

The risk discount rate β shows the percentage deviation of the certainty equivalent value from the expected value of a gamble. Again, fixing some outcome random variable at time period t and assuming $E\{x_t\} \neq 0$, we define β as

6 This follows the basic discounting formula in the literature (e.g. Price, 1993) which states that the present value of a future outcome x is given by the formula $PV_0(x_t) = x_t/(1+\alpha(t,p))$. Since our focus of interest is a risky future outcome, we can replace x by the certainty equivalence value of x. So, the formula can be represented as $PV_0(x_t) = CE(x_t)/(1+\alpha(t,p))$.

$$\beta(t,p) = \frac{E\{x_t\} - CE(x_t)}{E\{x_t\}} \tag{2}$$

where $CE(x_t)$ is the certainty equivalent of x_t at a specified time period t. It can be derived from the following formulations of expected utility and prospect theory, respectively:

$$U(CE(x_t)) = EU(x_t) = \sum_{i=1}^{n_t} p_{it} u(x_{it})$$

$$V(CE(x_t)) = E_\pi V(x_t) = \sum_{i=1}^{n_t} \pi(p_{it}) v(x_{it})$$

By rearranging equations (1) and (2), we obtain the following expressions:

$$PV_o(x_t) = \frac{CE(x_t)}{1+\alpha(t)}$$

$$CE(x_t) = (1-\beta)E\{x_t\}$$

Using the above expressions, we finally obtain the characterization of interest, expressing $PV_0(x_t)$ as the product of delay and risk discount rates, i.e.

$$PV_0(x_t) = \frac{1-\beta(t,p)}{1+\alpha(t,p)} \cdot E\{x_t\}$$

The present value $PV_0(x_t)$ is the fundamental treatment variable in the experiments. Using the present values reported by the subjects, the delay and uncertainty discount rates, α and β, will be elicited based on the formulation given above.

2.2 Predictions of the Models and Some Experimental Evidence

According to both the normative (DEU) and the proposed descriptive model (PT/IC), delay and uncertainty effects are independent and a simultaneous effect of delay and uncertainty can be obtained by combining each effect separately. Let's look at these predictions separately:

1. The delay effect does not depend on the uncertainty effect.

This implies that the future certain and risky prospects are discounted in the same fashion over time, i.e., $\alpha(t,p) = \alpha(t)\ for\ \forall\ p \in [0,1]$. However, this prediction is not supported by empirical evidence. One of the key studies that experimentally tested this behavioural account is by Keren and Roelofsma (1995). By conducting a series of choice-task experiments, they studied the relationship between delay and uncertainty. In one of the experiments, subjects in one treatment were to choose between an immediate and a delayed outcome, both yielding a certain payoff (receiving Fl. 100 now or in 4 weeks). 82% of the subjects in this treatment chose the immediate alternative, showing a pattern of delay discounting. In another treatment, the authors manipulated the choice pairs by adding a probabilistic dimension (receiving Fl. 100 with a probability of 50%). In this treatment, only 39% preferred the immediate reward to the delayed one. This suggests that risky prospects are discounted less heavily than certain ones. The authors argue that this may be due to the fact that uncertainty and delay discounting are not additive. More specifically, they claim that the implicit uncertainty discounting of a risky prospect may already be incorporated in delay discounting, resulting in a smaller discount rate for risky prospects than for certain ones.

Ahlbrecht and Weber (1997a) reported that certain events are discounted more heavily than uncertain ones in matching tasks. After eliciting the delay discount rates for a set of lotteries, the study tested for "certainty-risk asymmetry". One set of results show that in matching tasks, 55% of the subjects discounted a lottery of DM 120 with a probability of 0.99 less than they discount the certainty equivalent, when there was a delay of 6 months. When the delay was 24 months, this percentage increased to 73%. The authors link this to the fact that a simultaneous risk and delay discounting leads to a distraction from the need to discount, causing a subadditive discount rate. These results could not be confirmed in a choice task that was given to subjects. The authors explain this discrepancy between choice and matching tasks by the editing process:

individuals tend to cancel out common outcomes (in this case, common delays) in a choice task, whereas in a matching task, each prospect is evaluated separately.

2. *The uncertainty effect does not depend on the delay effect.*

This means that the risk preferences do not depend on whether the outcome is delayed or not. In other words, $\beta(t,p) = \beta(p)\ for\ \forall t \geq 0$. However, experimental evidence does not fit this model. In a study on the timing of uncertainty resolution, Ahlbrecht and Weber (1997b) experimentally tested the Kreps-Porteus dynamic choice model (1979). In their experiment, subjects were asked to state their preferences between different choice pairs and temporal lotteries. In one treatment the lotteries were to be resolved in the same day and in the other treatment, they were to be resolved two months in the future. The results show that subjects prefer delayed uncertainty resolution to immediate resolution. This would indicate that the rate of risk aversion declines over time. In other words, the value of a risky option increases as the resolution time is delayed. This finding is important for our study since it shows in an experimental setting that risk preferences *do* depend on the time period in which the uncertainty is resolved. If this is a general phenomenon, then "... any theory of probability discounting must incorporate delay, at least as a factor in framing of decisions".[7]

In a related study, Shelley (1994) used a managerial investment scenario to examine the changes in risk preferences for immediate and delayed gains and losses. In the experiment, the subjects were presented a set of lottery-type investment options, where the timings and uncertainties were manipulated, and they were asked to indicate their relative rankings. Looking at the implicit risk rates (i.e. uncertainty discounting), Shelley concluded that individuals who are risk (loss) averse in the present become more risk (loss) tolerant in the long run. These experimental findings suggest that the longer the delay of uncertainty resolution, the lower the value of the risky option becomes.

3. *A simultaneous effect of delay and uncertainty can be obtained by combining each effect separately.*

This prediction follows from the structure of the two models. In both DEU and PT/IC models, the delay and uncertainty effects are independent. Therefore, the models assume no interaction between these two effects. To

---

[7] Rachlin and Siegel (1994), pp. 172.

our knowledge, there is no experimental study that looks at the possible existence of an interaction effect between uncertainty and delay. However, the experimental studies mentioned above all highlight a common theme: Decisions concerning risky future prospects cannot be explained solely by time or risk preferences. In order to provide a sound descriptive model of intertemporal choice under uncertainty, we must examine the *interaction* between delay discounting (time preferences) and uncertainty discounting (risk preferences). The investigation of this interaction is the main objective of this paper.

3. EXPERIMENT

3.1 Experimental Design

This study focuses on the effects of delay and uncertainty in risky intertemporal decision-making. One such decision problem would be to choose between winning a sure amount of money today versus playing a lottery in the future. For instance, when evaluating a lottery that offers a 50% chance to win $500 six months from today, one's risk and time preferences play an important role in determining the exact present value of this gamble, that is how much one is willing to pay for it today. This study set out to examine the relationship between the risk and time preferences in such choice problems.

The experiment is designed to test three phenomena: the effect of delay on risk preferences, the effect of uncertainty on time preferences and the interaction of delay and uncertainty effects. Figure 3.1 shows a schematic presentation of the experimental design. The goal in the experiments is to examine the actual effects of delay and uncertainty. For this, we first separate these two different effects: the delay effect (whether the outcomes are evaluated in current time or in future) and the uncertainty effect (whether it's a risky outcome or not), as displayed by Figure 3.1. The paths α and α^* represent the effect of delay and the paths β and β^* show the effect of uncertainty. An illustrative example is given below to explain each effect. The example is based on the experiments that will be described later in the study.

Consider the case where an individual is indifferent between receiving $200 today and receiving $230 six months from today, we can calculate the effect of delay, α, as follows:

$$PV(x_t) = \frac{1}{1+\alpha} \cdot x_t \quad \Rightarrow \quad 200 = \frac{1}{1+\alpha} \cdot 230 \quad \Rightarrow \quad \alpha = 15\%$$

The upper left-hand cell in Figure 3.1 represents the sure amount received today (\$200), the lower left-hand cell represents the sure amount received in the future (\$230) and the path between these two cells represents the effect of delay α= 15%.

Now consider a lottery that offers a 50% chance to win \$420 today. If an individual is indifferent between playing this lottery and playing a lottery that offers a 50% chance to win \$500 six months from today, we can calculate the effect of delay, α^*, as follows:

$$PV(x_t) = \frac{1}{1+\alpha^*} \cdot x_t \quad \Rightarrow \quad 420 = \frac{1}{1+\alpha^*} \cdot 500 \quad \Rightarrow \quad \alpha^* \cong 19\%$$

The upper right-hand cell in Figure 3.1 represents the lottery resolved today, the lower right-hand cell represents the lottery resolved in the future and the path between these two cells represents the effect of delay α^*= 19%.

Figure 3.1 Delay and Uncertainty Effects

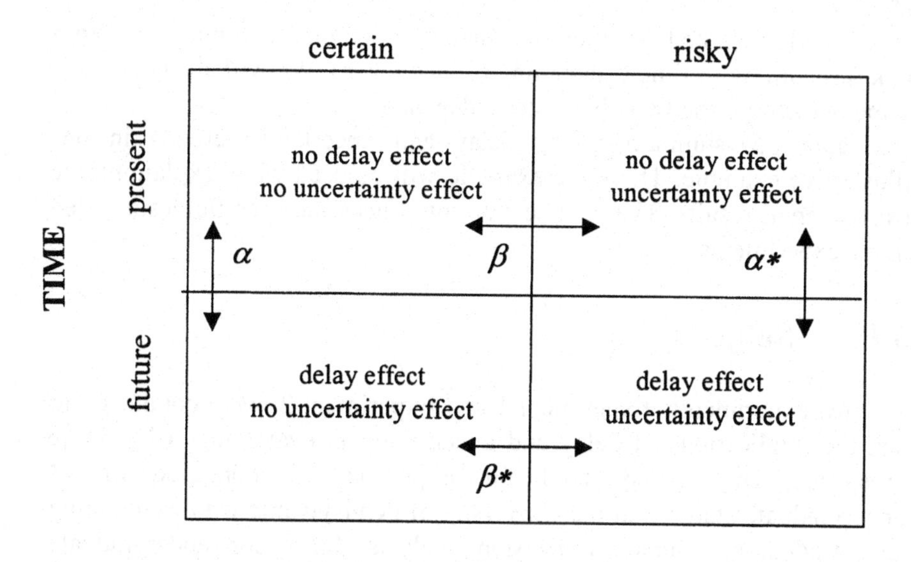

Next, let's suppose that this individual is indifferent between receiving a sure amount of $200 today and playing a lottery that offers a 50% chance to win $420 today. Noting that the expected value of the lottery is $210, we can calculate the effect of uncertainty β as

$$\beta = \frac{EV - CE}{EV} = \frac{210 - 200}{210} \cong 5\%$$

In this case, the upper right-hand cell in Figure 3.1 represents the lottery (50%, $420), the upper left-hand cell represents the sure amount ($200) and the path between these cells represents the effect of uncertainty, β= 5%.

Finally, if the individual is indifferent between receiving $230 six months from today and playing the lottery that offers a 50% chance of winning $500 six months from today, we can calculate the effect of uncertainty β^* as

$$\beta^* = \frac{EV - CE}{EV} = \frac{250 - 230}{250} = 8\%$$

In this case, the lower right-hand cell in Figure 3.1 represents the lottery (50%, $500) that is to be resolved in the future, the lower left-hand cell represents the sure amount ($230) that is to be received in the future and the path between these cells represents the effect of uncertainty, β^*= 8%.

It should be noted that it doesn't matter which direction one goes (from risk discounting to time discounting or vice versa). We will always get the same outcome if the two effects are independent.

Figure 3.2 summarizes the delay and uncertainty effects in our illustrative example. The same example will be used when explaining the experimental results. The next section introduces the specific design used in the experiments.

3.2 Subjects

Two experiments, Experiment 1 and Experiment 2, were conducted to test the implications of delay and uncertainty on evaluating risky future prospects. The subject pool for Experiment 1 consisted of 43 undergraduate students at the University of Pennsylvania who were taking an introductory course in decision analysis. Fifty-four undergraduate students at the summer school at the University of Pennsylvania participated as subjects in Experiment 2. Subjects were paid $5 each for taking part in the experiments. In each experiment, we presented the subjects a series of hypothetical decision problems on delayed and/or risky outcomes. The order of the decision problems was changed in each experiment to check for order effect and no order effects were found ($p<.05$).[8]

Figure 3.2 An Illustrative Example

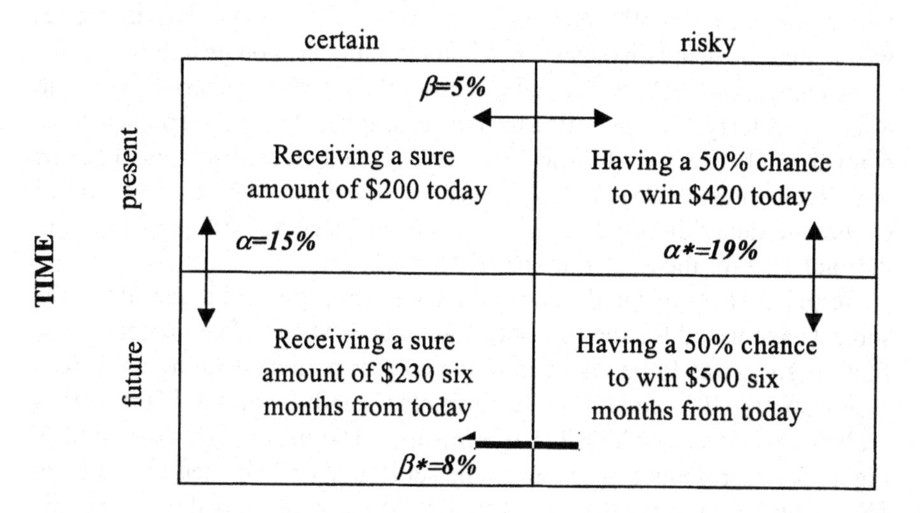

3.3 Parameters

The goal of Experiment 1 and Experiment 2 is to compare the effect of delay on risk preferences and the effect of uncertainty on time preferences. In order to measure the effect of delay and uncertainty, we will use the two quantities defined earlier: the delay discount rate $\alpha(t, p)$ and the risk discount rate $\beta(t, p)$.

3.4 Procedure

In the experiments, we followed the method previously described in order to derive $\alpha, \alpha^*, \beta, \beta^*$. In the first set of questions, we asked for the value for which the subjects would be indifferent between getting a

8 The questions for Experiment 1 and Experiment 2 can be obtained from the author.

delayed outcome or a present one. In the first question, the values were certain: (winning \$100 today versus winning \$X exactly one month from today) in Experiment 1 and (winning \$500 today versus winning \$X exactly six months from today) in Experiment 2. We derived α based on these figures.

In the second question, we asked the same question for a delayed **gamble**: (winning \$100 with a probability 40% today versus winning \$X with a probability 40% exactly one month from today) in Experiment 1 and (winning \$500 with a probability of 50% today versus winning \$X with a probability of 50% exactly six months from today) in Experiment 2. we elicited the delay discount rate α^* in the same fashion as before.

A comparison between α and α^* shows the effect of uncertainty on the effect of delay. In our illustrative example, this corresponds to a comparison between α=15% and α^*= 19%. Based on the findings of Keren and Roelosfma (1995) and Ahlbrecht and Weber (1997a), we would expect the delay discount rate for certain prospects to be higher than the discount rate for the risky ones ($\alpha^* < \alpha$).

In the next set of the decision problems, we presented a gamble to the subjects and asked for the certainty equivalence values. The gamble in the first question was to be resolved in the current period (winning \$100 with a probability 40% today) in Experiment 1 and (winning \$500 with a probability of 50% today) in Experiment 2. The upper right-hand cell of Figure 3.1 represents this gamble. The subjects then indicated the value of \$X for which they would be indifferent between playing the lottery and receiving a sure amount. In other words, \$X is the certainty equivalence value *(CE)* of the gamble. The upper left-hand cell in Figure 3.1 represents this value. Next we calculated β, the percentage deviation of the certainty equivalent from the expected value of the gamble.

The next gamble was identical to the first one, except for a delay in the timing of the resolution (winning \$100 with a probability of 40% **exactly one month from today** versus winning \$X **exactly one month from today**) in Experiment 1 and (winning \$500 with a probability of 50% **exactly six months from today** versus winning \$X **exactly six months from today**) in Experiment 2. The lower right-hand cell in Figure 3.1 represents the delayed gamble. The subjects were then asked to indicate their certainty equivalence values \$X for this distant gamble. The lower left-hand cell in Figure 3.1 represents this delayed sure amount. The percentage deviation of \$X from the expected value of the gamble, β^*, is calculated.

The effect of delay on risk preferences is examined by comparing the two values, β and β^*.In the example given in Figure 3.2, this corresponds to β=5% and β^*= 8%. If risk preferences were stationary over time, we would expect no difference between β and β^*. However, as we discussed before, studies like Ahlbrecht and Weber (1997b) and Shelley (1994) show that individuals who are risk- averse in the present become more risk-tolerant in the long run. Thus we expect to observe non-stationary risk preferences over time- specifically, less risk aversion- in the experiment ($\beta^* < \beta$).

Our third research question is to check if a simultaneous effect of delay and uncertainty can be described by combining each effect separately and to see whether the valuation of future risky outcomes is path-dependent. In order to evaluate a future risky outcome in terms of its present value, one may choose one of the following two paths: delay-uncertainty (path $\alpha\beta^*$) or uncertainty-delay (path $\alpha^*\beta$). In the delay-uncertainty sequence, one first discounts for the delay by applying a delay discount rate α and then for the uncertainty by applying a risky discount rate of β^*. In the uncertainty-delay path, an individual first accounts for the riskiness of the event by using a risk discount rate β and then discounts for the delay by a rate α^*.

To check for this path-dependency, in addition to the previous decision problems, we asked the individuals to state their present value for a future gamble (winning $500 with a probability of 50% exactly six months from today versus winning $X today) in Experiment 2. This is a comparison between the lower right-hand cell versus the upper left-hand cell in Figure 3.1. For example, let's assume the individual in our illustrative example states that he is indifferent between playing a lottery that offers a 50% chance to win $500 six months from today and receiving $200 today. To study how he evaluates this future gamble with respect to its present value, we compare this actual value with the present value calculated by using the discount rates (α, α^*, β, β^*) elicited in the previous questions. In our examples, the values for the elicited discount rates were α=15%, α^*=19%, β=5% and β^*=8%. Using these values, if we assume he discounts for the gamble first and then for the delay ($\alpha\beta^*$ path in Figure 3.1), his present value for the lottery would be $200:

$$PV(x_t) = \frac{1-\beta^*}{1+\alpha} \cdot E\{x_t\} = \frac{1-.08}{1+.15} \cdot 250 \cong 200$$

If he discounts for the delay first and then for the gamble ($\alpha^*\beta$ path), the present value would again be $200:

$$PV(x_t) = \frac{1-\beta}{1+\alpha^*} \cdot E\{x_t\} = \frac{1-.05}{1+.19} \cdot 250 \cong 200$$

Then we compare these two calculated values with the actual value ($200 in our example) to check if individuals discount future gambles in a sequential order.

Both normative and descriptive models, DEU and PT/IC predict that the two paths will yield the same net present value for a risky future outcome. The common assumption of these models is the separability of time and risk preferences. By studying the experimental results, we check whether the effects of delay and uncertainty are independent or whether a simultaneous presence of delay and uncertainty leads to a different evaluation of the risky future than predicted by a sequential process where the two effects are assumed to be independent.

4. RESULTS

4.1 The Effect of Delay on Risk Preferences

We first examine risk preferences over present and future gambles to see the effect of time on risk preferences (comparing paths β and β^* in Figure 3.1). In order to make this comparison, we compare two values: the sure amount of money, received now, that is equivalent to a gamble played now versus the sure amount of money, received later, that is equivalent to a gamble played later. The results show the future sure amount to be higher than the current one.

The descriptive statistics for the risk discount rates β and β^* are given in Table 4.1. Noting that a positive risk discount rate means risk-aversion, the mean values show that, on average, the subjects were risk-averse for both the present and the future gambles. In all cases, the discount rate was significantly higher than zero ($p < .01$ for all). The risk discount rate ranged between .07 and .20. The descriptive statistics also indicate that the subjects in Experiment 2 on average showed slightly less risk-aversion than the ones in Experiment 1 (.13 versus .20 for the present gamble and .07 versus .11 for the future one) but this difference is not statistically significant ($p<.98$).[9]

Our main focus in this study is a within-subject comparison between the current and future risk preferences. Comparing the mean values of β and β^* indicates that on average, subjects show significantly less risk-

9 Since the final reward is higher in Experiment 2, both the normative model DEU and the descriptive model PT/IC would predict the risk discount rate to be lower in Experiment 2 if the utility function $u(x)$ or the value function $v(x)$ has the property of decreasing risk aversion.

aversion for future lotteries. The risk discount rates are significantly higher for present lotteries than for their delayed equivalents in both of the experiments. Referring to our illustrative example, this would mean that a lottery that offers a 50% chance to win $500 is valued closer to its expected value when it is played six months from today, compared to the case when it is played in the current period.

Table 4.1: Mean Values (Standard Deviations) for Risk Discount Rates β and β*

	Experiment 1 n=43	Experiment 2 n=54
β	.20 (.11)	.13 (.07)
β*	.11 (.09)	.07 (.05)

Figure 4.1 shows the distribution of the difference between current and future risk discount rates for Experiments 1 and 2. Looking at the difference between β and β^* of the subject pool in the figure, we observe that almost all subjects became less risk-averse for the future gamble (moving from a mean value of β =.20 to β^* =.11 in Experiment 1 and from β =.13 to β^* =.07 in Experiment 2). Results show that there is a systematic difference between the two risk discount rates. We can reject the null hypotheses of $\beta - \beta^* = 0$ in both experiments ($p<.01$ for both cases). This means that risk attitudes are not stationary. Only 5% of the subjects in Experiment 1 and 7% in Experiment 2 kept their risk discount rates constant for both time horizons. Of the remaining participants, everybody in Experiment 1 and everyone except one subject in Experiment 2 decreased their rates, indicating a decline in risk-aversion.

4.2 The Effect of Uncertainty on Time Preferences

Another decision task in Experiments 1 and 2 was to choose between an immediate reward versus a delayed one in order to examine delay-discounting behaviour. The riskiness of the reward was manipulated in order to compare the discounting of certain and risky outcomes (comparing paths α and α^* in Figure 3.1). Based on the values given by the subjects, the delay discount rates, α and α^*, were elicited for both certain and risky delayed outcomes.

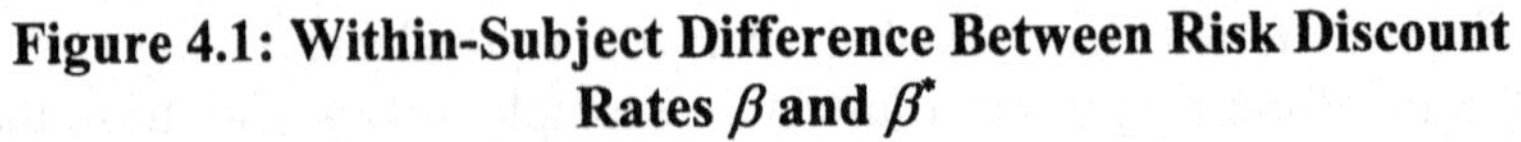

Figure 4.1: Within-Subject Difference Between Risk Discount Rates β and β^*

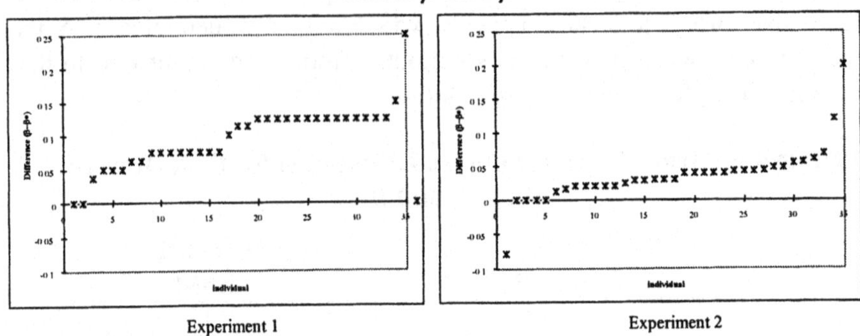

Comparing the delay discount rates shows that individuals discount the uncertain outcomes more than they do the certain outcomes. When subjects are asked what amount of money received with a .5 probability later is equivalent to \$500 received with a .5 probability now, the amount they state is higher than what they say when asked what amount of money at the same later time is equivalent to \$500 received for sure now. In our example given in Figure 3.2, this refers to the comparison between α= 15% and α^*=25%.

The mean values of the delay discount rates are presented in Table 4.2. The discount rates do not significantly differ in Experiments 1 and 2. Looking at the delay discount rates between certain and risky delayed outcomes, the hypothesis of equal group means can be rejected in both experiments ($p<.02$ for Experiment 1 and $p< .04$ for Experiment 2). The discount rates are lower for delayed risky prospects, compared to certain ones. This indicates that time preferences *do* depend on the riskiness of the outcome.

Table 4.2: Mean Values (Standard Deviations) for Delay Discount Rates α and α^*

	Experiment 1 n=43	Experiment 2 n=54
α	.21 (.05)	.23 (.06)
α^*	.25 (.04)	.27 (.08)

Figure 4.2 displays the distribution of the within-subject difference between discount rates for risky and certain outcomes for both experiments. The prediction from both models is that the delay discount does not depend on the riskiness of the outcome. The findings from both experiments suggest that most individuals treat the risky future differently than the certain future. Only 7% of the subjects in Experiment 1 and 9% in Experiment 2 use the same discount rates for certain and uncertain prospects. All of the remaining subjects, except for one in Experiment 1, discounted risky outcomes systematically more than they do the certain ones. This indicates that the riskiness of the outcome affects the evaluation of future prospects and the delay discount rate is not independent of uncertainty. More specifically, a delayed certain outcome is discounted less than a delayed risky one.

Figure 4.2: Within-Subject Difference Between Delay Discount Rates α and α*

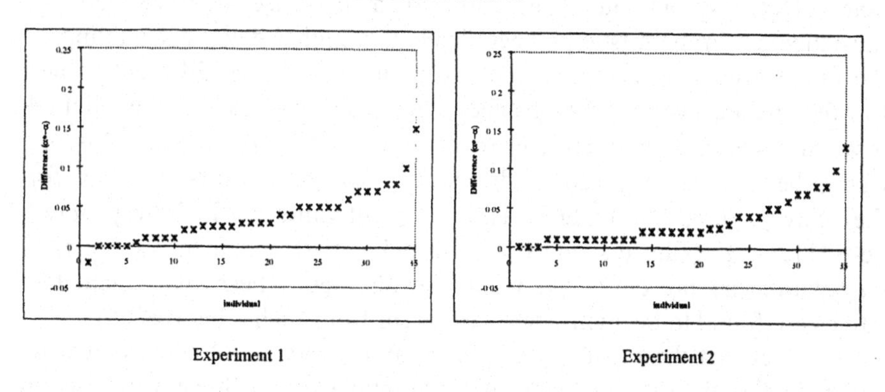

4.3 Interaction of Delay and Uncertainty Effects

The observations on the effects of delay and uncertainty suggest that there are two different paths to describe individuals' decision-making process for risky future prospects, which, in turn, yield different valuations. The first path is to determine the effect of uncertainty for a future gamble and then to account for delay discounting by applying the discount rate to the certainty equivalent (uncertainty-delay path $\alpha^*\beta$ in Figure 3.1). As we have calculated before, under this path, in our example in Figure 3.2, the present value of the lottery that offers a 50% chance to win $500 six months from today would be $200 since

$$PV(x_t) = \frac{1-.08}{1+.15} \cdot 250 = 200$$

An alternative way would be to determine the present value of the risky future prospect and then to discount for the uncertainty (delay-uncertainty path $\alpha\beta^*$ in Figure 3.1). Similarly, in our illustrative example, the present value of the future lottery under this path would again be $200:

$$PV(x_t) = \frac{1}{1+.25} \cdot 250 = 200$$

The models of risky intertemporal choice, DEU and PT/IC, predict that these two paths yield the same present value. Their common assumption is that the effects of delay and uncertainty are independent of each other. This means $\alpha=\alpha^*$ and $\beta=\beta*$ and as a result, the two ways to measure the present value of a risky prospect gives the same valuation.

The experimental findings, as discussed in the previous subsection, show that the effects of delay and uncertainty are not independent of each other. Delaying an outcome leads to a decrease in risk-aversion. Introducing a probabilistic prospect leads to a higher delay discounting of the future outcome. Therefore the two paths would give different values for the present value. For instance, the mean values for the elicited discount rates in Experiment 2 were α=23%, α^*=27%, β=13% and β^*=7%. Using these values, if a subject discount for the gamble first and then for the delay ($\alpha\beta^*$ path in Figure 3.1), his present value for the lottery would be $211.69. If (s)he discounts for the delay first and then for the risk ($\alpha^*\beta$ path), the present value would be $216.60. These two calculated values are found to be significantly different from each other ($p<.05$).

The fact that these paths yield different answers is not consistent with either of the theories of risky intertemporal choice. Since we observe different values for different paths in an experimental setting, the next question would be which path describes the decision process more accurately. If neither has a satisfactory descriptive power, then we would investigate the existence and direction of an interaction between delay and uncertainty effects, which cause a different valuation of future gambles than predicted by combining each effect separately.

In Experiment 2, we asked the subjects an additional question to see which, if any, of these paths explain their valuation. We asked them to state a certain present value for which they would be indifferent in playing a lottery (.5, $500; .5, $0) six months from today and receiving the sure amount immediately. This question was asked in addition to the ones in Experiments 1 and 2. For each individual, we compared the actual value with the two calculated values under the paths $\alpha\beta^*$ and $\alpha^*\beta$. By doing this, we check if delay-uncertainty or uncertainty-delay paths explain the behaviour better and whether a simultaneous presence of delay and uncertainty leads to a different evaluation of the risky future than predicted by a sequential process.

We found that when the effect of delay and uncertainty are elicited separately and used to calculate the present value of a future gamble, both of the current calculated values are much higher than the actual current values given by the subjects. In other words, if individuals had discounted a risky future prospect sequentially ($\alpha\beta^*$ or $\alpha^*\beta$), they would attach a higher value to the prospect than they actually do. Therefore, there is evidence that there exists an interaction effect due to a simultaneous existence of delay and uncertainty. If it were only the future certain outcome or the present risky outcome whose values were being evaluated in a distorted fashion, then one of the $\alpha\beta^*$ or $\alpha^*\beta$ paths in Figure 3.1 would have predicted the valuation of the future risky outcome.

Table 4.3: Mean Values (Standard Deviations) for Present Values (n=54)

Uncertainty-Delay Path	Delay-Uncertainty Path	Actual Value
211.69 (9.71)	216.60 (8.47)	203.54 (10.24)

Table 4.3 presents the descriptive statistics for the present values in Experiment 2. The first column shows the mean value for the derived present value of the future lottery based on the uncertainty-delay path. The second column gives the derived value for the delay-uncertainty path. The last column is the actual present value given by the subjects. All three of the mean values are found to be significantly different from each other ($p<.05$). On average, uncertainty-delay path predicts $PV_0(X_t)$ to be \$211.69 and delay-uncertainty path predicts it to be \$216.60, whereas the actual mean value of $PV_0(X_t)$ is found to be \$203.54. The actual value is significantly lower than the predicted values ($p < .01$). This suggests that (1) there *is* an interaction of delay and uncertainty and (2) this interaction leads to an even higher discounting of the risky future than either effect would if operating individually. A speculative discussion on some possible reasons for these results is presented in the next section.

Figure 4.3 displays the within-subject distribution of actual present values of X_{tp}, along with the values predicted under uncertainty-delay and delay-uncertainty paths. For most of the subjects (93% of the subject group), the actual present value of a risky future outcome is lower than those predicted by *either* path. Neither of the paths explains the discounting of future risky prospects completely. There is only one subject whose present value matches with the one predicted by uncertainty-delay path (with a present value of $196) and one subject whose present value can be explained by delay-uncertainty path (with a present value of $200). For the rest of the subjects, the sequential nature of discounting risky future does not apply. This indicates that a simultaneous presence of delay and uncertainty leads to a different evaluation of the risky future than predicted by either of the sequential processes. In the next section, we elaborate more on the reasons and implications of this result on the intertemporal risky models of DEU and PT/IC, discussed in Section 2

Figure 4 .3: Distribution of $PV(X_{tp})$ in Experiment 2

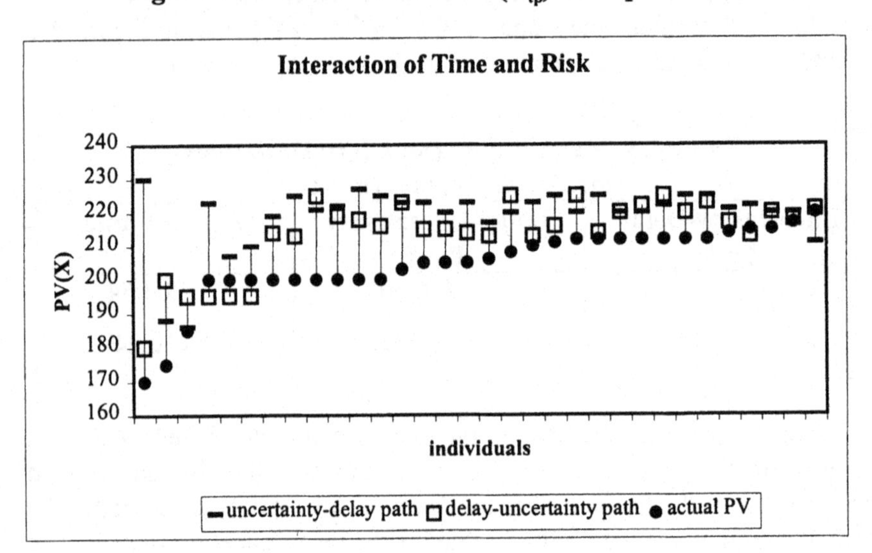

5. SUMMARY AND CONCLUSIONS

The findings from Experiment 1 and Experiment 2 suggest that the decision process for evaluating risky future prospects is not well-modeled by either of the models, DEU or PT/IC, which assume independent time and risk preferences. The findings can be summarized as follows:

The effect of delay on risk preferences: Postponing a lottery to the future increases the future certainty equivalence value of the lottery. So, a lottery offering $500 with probability .5 is valued closer to expected value (in terms of the certainty equivalence to be received in six months) when it is played six months from today, compared to the case when it is played in the current period (and valued in terms of current certainty equivalence). In order to make this comparison, we compared two values: the sure amount of money, received now, that is equivalent to a gamble played now versus the sure amount of money, received later, that is equivalent to a gamble played later. We found the future sure amount to be higher than the current one. This is a comparison of the upper two cells with the bottom two cells in Figure 3.1.

The results show that, in the gain domain, there is a systematic reduction in individuals' risk aversion when the outcome is delayed ($\beta > \beta^*$). In other words, referring to our illustrative example, the lottery that offers a 50% chance of winning $500 is valued more in the future, relative to its discounted expected value, than in the current period, relative to its expected value.

These results are consistent with Ahlbrecht and Weber's (1997b) experimental findings that suggest that individuals become less risk-averse as the uncertainty resolution is delayed. As we have discussed in Section 2, Ahlbrecht and Weber's results show that subjects prefer delayed uncertainty resolution to current resolution. This implies that the subjects tended to be less risk-averse for delayed lotteries and the rate of risk aversion declines over time. Our experimental results support this finding.

One possible reason for a decline in risk-aversion over time is the anticipation effect. The anticipation effect (Loewenstein, 1987) describes additional utility (disutility) associated with the delayed consumption of a desirable (undesirable) good. Elster and Loewenstein (1992) define *savoring* as the positive utility derived from anticipating a desirable outcome and *dread* as the disutility derived from anticipating an undesirable outcome. In the gain domain, anticipation may be a source of positive utility, causing individuals to experience the hedonic utility of the future prospect in the current period. If we assume that the anticipation of winning a gamble overweighs the anticipation of winning a sure amount, this extra utility of a possible future gain might be one reason why individuals are less risk-averse for future gambles, compared to current gambles.

The effect of uncertainty on time preferences: Individuals discount the uncertain outcomes more than they do the certain outcomes. A comparison between α and α^* shows the effect of uncertainty on the effect of delay. The findings show that individuals discount the uncertain gambles more than they do the certain outcomes. In other words, the delay discount rate α increases as the outcome becomes probabilistic ($\alpha^* > \alpha$). In our example given in Figure 3.2, this corresponds to α^*=25% and a=15%.

This result is not consistent with the findings of Keren and Roelofsma (1995) and Ahlbrecht and Weber (1997a). The experimental findings from the first study suggest that in a choice task, risky outcomes are discounted less than certain ones. The authors claim that the implicit discounting for uncertainty overlaps with delay discounting and therefore, it reduces the final level of delay discount rate. Ahlbrecht and Weber (1997a) obtained similar results for a matching task, but not for a choice task. Our results show, contrary to these findings, that the presence of riskiness aggravates the delay discounting process.

One reason for this difference in the results might be due to the differences in the experimental designs. Keren and Roelofsma's study was a between-subject choice experiment whereas the current study imposes a within-subject matching task. Ahlbrecht and Weber's study had both choice and matching tasks and it imposed a within-subject design like ours. The difference between the two experiments is the choice of matching task questions. The previous study compared the discount rate for the present value of a future certainty-equivalent and the discount rate for the present value of a future lottery. The authors propose that discounting for risk already incorporates discounting for delay and therefore, risky outcomes are discounted less than certain ones. In the current experiment, we compare the present value of a future certainty equivalent with the present value of a delayed *certain* outcome. Therefore, we compare the delay discount rates for the risky outcomes with the certain ones, without any effect of risk preferences. Within this design, we showed that risky outcomes are discounted more heavily than certain ones. This difference in the matching tasks may have led to different results. However, the reason for this sensitivity to experimental design is not yet clear. Further research is needed in order to verify the direction of delay discounting when shifting from certain prospects to risky ones.

One possible reason for obtaining higher discount rates for risky future prospects might again be due to the anticipation effect (Loewenstein, 1987). The anticipation effect generates an additional utility associated with the delayed consumption of a good. This positive utility is predicted to be higher for future certain outcomes than for gambles, as mentioned by Elster and Loewenstein (1992, pp. 227). One possible explanation offered by the authors is that individuals may have a more difficult time in predicting the hedonic utility of a future risky event than a future certain one. This difficulty would predict a lower anticipation effect for risky prospects, consistent with what was observed in our experiment: Future prospects that have certain outcomes are thus discounted systematically lower than risky prospects (for the same amount of positive monetary outcome.

<u>The interaction of delay and uncertainty effects</u>: A key finding of this study is that a simultaneous presence of delay and uncertainty leads to a lower valuation of a future risky prospect than predicted by each effect separately. This means that the present value of a future gamble (winning $500 with a probability of 50% exactly six months from today versus

winning $X today) is lower than the values predicted by each effect separately.

One possible explanation for the superadditivity in discounting future gambles can be a "temporal uncertainty avoidance". Hofstede (1990) defines "uncertainty avoidance" as a preference for outcomes with no or little uncertainty over outcomes with a probabilistic dimension. We can speculate that for temporally extended outcomes, this uncertainty avoidance is even higher. This would cause further underweighting of future gambles. It would be interesting to study the factors which may affect this temporal uncertainty avoidance, such as the magnitude of wins and losses, the precise likelihood of the outcomes and the remoteness of the uncertainty resolution. It may also be worthwhile to focus on the modeling of risky intertemporal choice. Since linearly additive variables for delay and uncertainty do not capture the interaction between these two effects, an alternative model of intertemporal choice under uncertainty should be constructed, including an interaction variable between delay and uncertainty effects. Another dimension of the problem is the utility (value) function. Both DEU and PT/IC models assume this function is stationary. In other words, the value of a risky prospect is the same regardless of the time period. Relaxing this restrictive assumption might actually show that the value (utility) of a risky prospect changes with respect to the specific time period we look at. Examining this possibility is another direction for future research.

The interaction between delay and uncertainty effects decreases the present value of risky future outcomes significantly over what would be predicted by using a non-stationary discount rate and risk preferences combined. These behavioural findings suggest a systematic undervaluation of future risky prospects. This simultaneous effect of delay and uncertainty provides an explanation for the low valuations of risky future prospects in many different domains, including earthquake mitigation. This phenomenon may explain why individuals do not adopt risk mitigation measures against natural hazard risks associated with high probability-low consequence events, compared to investing in energy-saving appliances which promise a certain gain in the future (See Kunreuther *et al.*, 1998).

In this paper, we designed an experiment to study intertemporal choice under uncertainty. Our main finding is that the interaction between time and risk preferences leads to a high discounting of future risky prospects. For future research, it is worthwhile to explore possible heuristics that lead to this undervaluation of risky future prospects in different domains, especially in risk mitigation investments and to contemplate on alternative models of risky intertemporal choice. By understanding the way information is processed and the limitations on thought processes, better policies may be designed to overcome myopia and similar biases in intertemporal decision-making under uncertainty.

REFERENCES

Ahlbrecht M. and M. Weber. 1997a. "An Empirical Study on Intertemporal Decision Making Under Risk," *Management Science* 43: 813-26

Ahlbrecht M. and M. Weber. 1997b. "Preference for Gradual Resolution of Uncertainty," *Theory and Decision* 43: 167-185.

Ainslie G. 1991. "Derivation of Rational Economic Behaviour from Hyperbolic Discount Curves," *American Economic Review Papers and Proceedings* 81: 334-340.

Benzion U., A. Rapaport and J. Yagil. 1989. "Discount Rates Inferred from Decisions". *Management Science* 35: 270-284.

Elster J. and G. Loewenstein. 1992. "Utility from Memory and Anticipation." In *Choice over Time*, edited by G.Loewenstein and J.Elster. Russell Sage Publications.

Hofstede, G. 1990. Culture's Consequences: International Differences in Work-Related Values. Beverly Hills, CA: Sage.

Kahneman D. and A. Tversky. 1979. "Prospect Theory: An Analysis of Decision Under Risk," *Econometrica* 47 (2): 263-291.

Keren, G. and P. Roelofsma. 1995. "Immediacy and Certainty in Intertemporal Choice," *Organizational Behaviour and Human Decision Processes* 63 (3): 287-297.

Koopmans, T.C. 1960. "Stationary Ordinal Utility and Impatience," *Econometrica* 28: 207-309.

Kreps, D.M. and E.L. Porteus. 1979. "Temporal von Neumann-Morgenstern and Induced Preferences." *Journal of Economic Theory* 20: 81-109.

Kunreuther, H., A.Onculer, and P. Slovic. 1998. "Time Insensitivity for Protective Investments," *Journal of Risk and Uncertainty* 16: 279-299.

Loewenstein G. 1987 "Anticipation and the Valuation of Delayed Consumption," *Economic Journal* 97: 667-684.

Loewenstein G. and D. Prelec. 1991. "Decision-Making Over Time and Under Uncertainty: A Common Approach". *Management Science* 37: 770-786.

Loewenstein G. and D. Prelec. 1992. "Anomalies in Intertemporal Choice: Evidence and an Interpretation." In *Choice over Time*, edited by G.Loewenstein and J.Elster. Russell Sage Publications.

Lucas R.E. 1978. "Asset Prices in an Exchange Economy," *Econometrica* 46 (6): 1429-1445.

Mehra R. and E.C. Prescott. 1985. "The Equity Premium: A Puzzle," *Journal of Monetary Economics* 15: 145-161.

Price C. 1993. *Time, Discounting and Value.* Oxford: Blackwell Publishers.

Rachlin H., A.W. Logue, J. Gibbon and M. Frankel. 1986 "Cognition and Behavior in Studies of Choice," *Psychological Review* 93: 33-45.

Rachlin H. and E. Siegel. 1994. "Temporal Patterning in Probabilistic Choice," *Organizational Behavior and Human Decision Processes* 59: 161-176.

Samuelson P. 1937. "A Note on Measurement of Utility," *Review of Economic Studies* 4:2.

Shelley M.K. 1994 "Gain/Loss Asymmetry in Risky Intertemporal Choice," *Organizational Behavior and Human Decision Processes* 59: 124-159.

Thaler R. 1981. "Some Empirical Evidence on Dynamic Inconsistency," *Economics Letters* 8: 201-207.

Tversky A. and D. Kahneman. 1992. "Advances in Prospect Theory: Cumulative Representation of Uncertainty," *Journal of Risk and Uncertainty* 5 (4): 297-323.

Von Neumann J. and O. Morgenstern. 1944. *Theory of Games and Economic Behavior*. Princeton University Press.

7

A Social Decision Analysis of the Earthquake Safety Problem:

The Case of Existing Los Angeles Buildings *†

Rakesh Kumar Sarin
The Anderson School at UCLA

In this paper, we propose a framework for conducting a decision analysis for a societal problem such as earthquake safety. The application deals with the formulation and evaluation of alternative policies for the seismic safety problem faced by the city of Los Angeles with regard to its old masonry buildings. A social decision analysis compares the costs and benefits of the alternative policies from the viewpoints of the impacted constituents. The emphasis is on identifying acceptable policy that considers the interests of the impacted constituents and provides incentives for their cooperation. Alternatives ranging from strict regulation to free market are examined. In order to evaluate the trade-offs between additional cost and savings in lives, a direct willingness-to-pay and an economic approach, based on property value differential, are used. Recommendations range from strict regulation for the residential and critical buildings (schools, hospitals, fire stations, etc.) to simply informing the occupants (in the case of commercial and industrial buildings) of the risks involved.

* It is a pleasure to acknowledge the helpful input of Earl Schwarz, Senior Structural Engineer, Department of Buildings and Safety, Los Angeles. The helpful suggestions of Ralph Keeney and David Okrent are thankfully acknowledged. Jacob Szabo served as a research assistant and provided valuable help in data collection. The work on this paper was supported by an NSF Grant 79-10804.

† This paper is essentially the same paper as published in *Risk Analysis*, 3(1), 1983. An appendix to update the status of unreinforced masonry buildings has been added.

P.R. Kleindorfer and M.R. Sertel (eds.), Mitigation and Financing of Seismic Risks, 129–157.

1. INTRODUCTION

A cursory look at any of the annual volumes of the *New York Times Index* under the heading "Earthquakes" will reveal their devastating nature. The cruel impact of earthquakes on mankind was most recently dramatized when 650,000 lives were lost in Tang-Shan, China (July 28, 1976). Earthquakes occur with alarming regularity. Every few years, earthquakes devastate cities inflicting heavy losses of life and high injury rates. In those lucky years when the impact of earthquakes is less severe, it is because these earthquakes are centered in areas of low population density, rather than that fewer earthquakes occur. With the population growth leading to the development of large cities, the potential for great earthquake destruction increases every year.

Residents of many cities and counties in the United States, particularly in California, face a potential risk of death and injury by partial or complete collapse of buildings in the event of an earthquake. It is possible to reduce these risks by raising the standards for structural seismic resistance of the buildings. This, however, would require costly modifications by the owners of private buildings or by the city or state government, in the cases of public buildings.

The problem of selecting an appropriate upgrading standard for seismic resistance of the existing buildings is difficult for the following reasons:

1. There is significant uncertainty about the relationship between building standards and resulting deaths and injuries since this depends on the location and intensity of the earthquake, the extent of damage to the property, and the number of occupants at the time of damage.
2. It involves value judgment to determine an acceptable trade-off between the cost of upgrading and the reduction in deaths and injuries.
3. There are several different constituents, for example the owners of the buildings, the renters, the planners, the policymakers, and the public-at-large, who are impacted differently and thus may have differing viewpoints.

It is therefore clear that the selection of a policy for the earthquake safety problem cannot be left to an intuitive determination. We use the "divide and conquer" strategy of decision analysis (Raiffa 1968) to decompose the problem into several components. These components are then analyzed and integrated using scientific facts, professional judgments of the experts, and the value trade-offs of the impacted constituents.

This approach is applied to the design and evaluation of alternative policies for the earthquake safety problem of the city of Los Angeles. The framework is, however, general enough to be useful to other cities and for other problems involving risks to human health and lives. Since the emphasis here is on a real world application, we have omitted methodological details and have presented the ideas in a non-technical style.

In Section 2, a brief background of the earthquake safety problem of the city of Los Angeles is given. An overview of our approach is provided in Section 3. The results of a detailed social decision analysis are presented in Section 4. These results represent an order-of-magnitude analysis. The city of Los Angeles was considering an ordinance for earthquake hazard reduction at the time that this study was commenced. This ordinance was passed by the city council on January 7, 1981. In Section 5, we provide our recommendations and compare these with the provisions of the city ordinance. Finally, conclusions are given in Section 6.

2. BACKGROUND OF THE EARTHQUAKE SAFETY PROBLEM OF THE CITY OF LOS ANGELES

Los Angeles, like many other cities in California and in the nation, has a large number of existing earthquake-hazardous buildings. These buildings were built before earthquake standards were incorporated into the building codes. In case of a major earthquake, these buildings are most susceptible to collapse, causing deaths and injuries among the occupants. This paper specifically deals with the unreinforced masonry buildings that were built before 1933, prior to code requirements designed to withstand earthquakes. We have chosen to focus on these buildings for two reasons:

1. These buildings pose the greatest hazard to human life and property (*Science* 1982, 385-387; U.S. Department of Commerce 1973).
2. Sufficient information on the types, uses, occupancy, and so forth for these buildings has been complied by the city. Similar information for the other buildings is unavailable at this time.

We now provide some factual information on these pre-1933 buildings. Detailed information is contained in Sarin (1982).

Los Angeles has approximately 7,863 unreinforced masonry buildings that were built before 1933; approximately 1 million people live or work in them. Excluded from these figures are detached dwellings and detached apartment houses containing fewer than five units. Approximately 10% of the buildings are residential. According to the occupancy load and the use of the buildings, all buildings are classified by the city in four risk classes: essential, high risk, medium risk, and low risk:

1. *Class I: Essential Buildings* – Those structures or buildings that are to be used for emergency purposes after an earthquake, in order to preserve the peace, health, and safety of the general public.
2. *Class II: High-Risk Buildings* – Any building other than an essential building having an occupant load of 100 occupants or more, wherein the occupancy is used for its intended purpose for more than 20 hr/week.
3. *Class III: Medium-Risk Buildings* – Any building having an occupant load of 20 or more occupants that is not classified as Class I or Class II.
4. *Class IV: Low-Risk Buildings* – Any building, other than Class I, having an occupant load of less than 20 occupants.

The cost of strengthening these buildings against seismic forces is estimated to be $1 billion. While the costs for upgrading the buildings are large, the risks to property and human life are also significant since many experts believe that a great California earthquake is imminent (Bolt 1978; Bolt 1978). Further, these relatively large risks of deaths and injuries are faced by an identifiable segment of the population, occupants of the old buildings. After years of deliberations, the Los Angeles City Council passed an ordinance (Ordinance No. 154,807) on January 7, 1981 that will require rehabilitation of these buildings to some specified standards.

3. OVERVIEW OF THE APPROACH

Decision analysis has been used for a wide variety of problems during the past decade (Keeney 1982). A social decision analysis is essentially a decision analysis of a societal problem in which the interests and preferences (often conflicting) of different members of the society are considered in the choice of a preferred alternative. In the following, we discuss some important features of a social decision analysis in the context of the earthquake safety problem of the city of Los Angeles.

3.1 Decision Alternatives

In a social decision analysis, a full range of regulatory as well as freemarket decision alternatives must be considered. For the earthquake safety problem, at one extreme are the strict regulations for the design and construction of all buildings that are based on the best available information and evaluation by the government. The other extreme is to treat safety as an economic commodity and let the freemarket mechanism along with professional codes of practice and existing liability laws determine the acceptable levels of standards for each type of building. The former policy option suffers from the difficulty of monitoring and enforcement, whereas the latter presupposes that the individual members of the society can assess and evaluate possible risks (through freemarket mechanisms such as insurance companies or building inspection companies), and that the total cost of information dissemination will be lower than the cost of regulation. Besides the monetary costs, there are ethical arguments in favor and against each of these extreme positions.

Intermediate policy options include some form of government intervention in specifying seismic resistance requirements for critical facilities (schools, hospitals, etc.), while *providing information* to the owners and the occupants about earthquake hazard and mitigation alternatives for the existing buildings. In evaluating alternative risk-management policies for Los Angeles, we consider several of these policy options.

It is surprising that in many decision situations, the full range of alternatives is not considered. For example, if we assume that the recommendations in the Los Angeles city ordinances are optimal in some sense, still the optimality is with respect to the subset of alternatives that fall within the regulatory options. A creative generation of the alternatives is a very important step in the social decision analysis.

3.2 Impact on the Constituents

In a social decision analysis, the consequences of a chosen alternative are felt differently by different constituents. Further, the final outcome depends on the actions of the constituents in reaction to the chosen alternative. It is therefore imperative that the impacts on different constituents should be considered explicitly in the evaluation of alternatives. We show that an *interactive decision tree* can be used for this purpose. The key feature of the interactive decision tree is that for each alternative the reactions of the impacted parties (such as the owners of the buildings in our case study) are treated as explicit decisions that influence the final outcome.

In the earthquake safety problem, a decision by the city to rehabilitate the old buildings will impact upon several constituents. The directly affected constituents are the renters and the owners. Owners of the buildings will have to pay the cost of upgrading or share it with the city if some financial incentives are offered. They would, however, receive benefits in reduced property damage if an earthquake occurs, a possible appreciation in the value of the building, reduced liability in case a renter gets injured or killed, and possibly higher future income from rents. An owner will prefer the upgrading if the benefits are higher than the cost. If the owner is unaware of the benefits, since much of these benefits occur in the future and are uncertain, proper information could entice him to undertake upgrading of his building. It seems that the owners are quite resistant to upgrading the buildings so it is possible that they do not perceive the benefits to be greater than the costs.

The renters of the buildings are another affected group. Strengthening of a building clearly makes the building safer to live in, but the rents might also increase, and some renters may have to leave the building temporarily during the construction phase. A majority of the renters of the old residential buildings are from lower economic classes. These people can ill afford to pay substantially higher rents. It is therefore unclear without an explicit examination of the costs and benefits whether upgrading is attractive from the renters' viewpoint.

Policymakers and planners constitute the third group who is affected, though indirectly, by the city's action. If the city requires costly upgrading, the sentiments of the owners run against them. The letter of one owner, Robert M. Lawson, to Councilman John Ferraro, with regard to the city ordinance requiring upgrading of the buildings is representative of how a majority of the owners feel about upgrading, "the passage of such an ordinance would destroy one of the principal remaining assets in my family. This is a poor reward for 53 years of highly productive participation in the economic growth and development of Los Angeles." If the city leaves these buildings alone and if a major earthquake does destroy them, causing deaths and injuries to the occupants, then the policymakers will be held responsible for their inaction.

Finally, the public-at-large is also an affected party. Since the group that suffers the most damage in case of an earthquake is identifiable a priori, the members of the society who do not live or work in these buildings would be willing to pay some amount for the safety of the occupants of these buildings. The benevolent considerations become especially important if the public perceives that the residents of the hazardous buildings are unfairly treated because of their age, income, or other social conditions.

3.3 Objectives of the Decision Problem

In order to formulate and evaluate alternative policies, the objectives of the decision problem should be clearly specified. Often, there are multiple conflicting objectives. For example, one objective in the earthquake safety problem is to reduce deaths and injuries, while the other is to reduce the cost of rehabilitating unsafe buildings. These objectives cannot be met simultaneously. The welfare of the landlords may be in conflict with the welfare of the tenants, and so on. It is therefore important to identify the attributes relevant to all constituents that may be affected by a policy. A discussion of a hierarchical approach for identifying attributes of a decision problem is provided in Keeney and Raiffa (1976).

3.4 Quantification of Consequences

The eventual consequences of a chosen alternative are uncertain in a complex societal problem such as earthquake safety. Some consequences can be predicted with reasonable accuracy. In general, however, the consequences must be described in the form of a probability distribution. The probability distributions over the consequences are often quantified using a combination of past data, sample scientific and engineering studies, and experts' professional judgments. Appropriate methods for quantifying probabilities (Spetzler and von Holstein 1975) and models that simplify probability assessments are helpful in this assessment task. In most, if not all, social decision analysis the professional estimates are used in conjunction with data for assessing the consequences. This is done on the premise that the affected individuals often do not have sufficient information to provide such estimates. As demonstrated by several cognitive psychologists (Slovic, Fischoff and Lichtenstein 1980), an individual's estimates based on his perception may be considerably different than the so-called scientific estimates. Unfortunately, however, the acceptability of a chosen policy alternative by the affected parties depends crucially on these perceived costs and benefits (reduced risks). If the scientific estimates are different from the public perception, efforts should be made to better inform the public. In some situations, dissemination of the scientific information may narrow the gap between the perceived and the scientific estimates. Market research studies (Green and Srinivasan 1978) could also be helpful in quantifying public perception and an analysis could be conducted based both on the scientific information and the information obtained by the market research. In some situations, a combination of the objective scientific data, experts' opinions, and the market research data may be used for analysis. A sensitivity analysis could point out the need for refining or reconciling various estimates.

3.5 Quantifying Preferences

In a social decision analysis, the preferences of the affected constituents must be considered in determining the trade-offs between the attributes. In the earthquake safety problem, the key trade-off is between additional cost and safety (measured in terms of reduced injuries and deaths). We considered three approaches for quantifying the preferences in our application.

The first approach assumed that the trade-off between cost and safety is in accordance with the empirical evidence on how much society does indeed pay for reducing similar risks. Essentially, in this approach the information from published sources (Bailey 1980) was used to quantify cost/safety trade-offs.

In the second approach, a small number of the occupants of the buildings were interviewed directly to determine their willingness to pay for safety. The interview data was then used to quantify cost/safety trade-offs.

In the third approach, it was considered that the cost/safety trade-offs should be inferred from the property value differential between earthquake-safe and earthquake-unsafe buildings. This approach, called the "hedonic price" approach, has been extensively developed in the economics literature (Brookshire et al. 1982).

Each of the three approaches offers some advantages and some disadvantages. We feel that in a social decision analysis several approaches to quantify key trade-offs should be used.

3.6 Evaluation and Policy Formulation

The objective of evaluation is to compare the costs and benefits of the alternative policies and select the policy that is most preferred with respect to the preferences of the affected constituents. Unfortunately, however, the preferred policy may be different for different constituents. We consider that the first step in evaluation is to compare total costs with total benefits regardless of to whom they accrue. This analysis ignores the distributional aspects of costs and benefits, for example, some groups may experience a relatively larger share of costs or benefits. The second step in the evaluation is to compare the costs and benefits from the viewpoint of various affected constituents, for example, tenants, owners, planners, and policymakers in our case study. Based on these evaluations, an acceptable policy can be formulated. A critical issue in formulating a policy is to ensure that it can be implemented and the enforcement is possible within the means of the concerned agency (city or state government). For example, a report by Stanley Scott (1979) points out that "a crucial weak point in seismic safety policy is the enforcement of seismic design regulations."

An acceptable policy should address several real world institutional, socioeconomic, and political constraints. Public acceptability and institutional mechanisms to carry out a policy often decide the success or the failure of a risk-management policy. An explicit consideration of all possible hurdles that could impede the implementation of a policy must be well thought out. Full ventilation of the diverse and differing opinions and a full awareness of the potential problems before a policy is finally adopted are requisites for its success. Needless to say, all discontent cannot be eliminated and all affected parties cannot be fully satisfied. An understanding of their concerns would greatly improve the design of a policy.

A word of caution is warranted here. A good policy is not the one that attempts to incorporate every single concern of the time. But a good policy recognizes what can be changed and attempts to do so if such a change is in the greater welfare of the society; simultaneously, it recognizes the boundaries within which it must operate.

In the earthquake safety problem we show how an acceptable policy that considers the interests of the impacted constituents and provides incentives for their cooperation can be formulated. The formulation of such a policy is based on the analysis of costs and benefits that accrue to various affected constituents.

A technical point that may be relevant in some social decision analysis warrants some attention. In our study, all analyses assume an additive linear preference function. If the range of consequences on each attribute is relatively small, this form is reasonable. Moreover, if one considers an array of policy contexts in which the same attributes, for example cost and safety, are impacted (e.g., fire safety, storage of chemicals, transportation of hazardous material, etc.) then for *any one policy* context, such as earthquake safety, the range of consequences is often not large. We favor the use of the additive linear form since its informational requirements are small and it is easy to interpret. Keeney and Raiffa (1976) describe more complex forms that permit preference dependencies among attributes. Keeney (1981) shows how by redefining attributes, the additive linear form becomes appropriate even for those cases where preferences among attributes exhibit dependencies.

4. ANALYSIS AND RESULTS FOR THE EARTHQUAKE SAFETY PROBLEM

In this section, we present the analysis and results for the earthquake safety problem of the city of Los Angeles. In Section 4.1, the structure of the decision problem is described. An estimation of economic and human health consequences is given in Section 4.2. A comparison of the costs and benefits of alternative policies is made in Section 4.3.

4.1 Structure of the Decision Problem

4.1.1 Alternatives

Buildings are divided into four risk classes: Essential (Class I), high risk (Class II), medium risk (Class III), and low risk (Class IV). In each risk class, several upgrading alternatives can be undertaken. These alternatives are: leave the buildings to their present Masonry C status, upgrade to Masonry B standard, upgrade to Masonry A standard, and upgrade to Today's Standard. (Construction qualities A, B, and C refer to the degree of earthquake-resistance provided.)

Construction quality A includes good workmanship, mortar, and design; has reinforcement, especially lateral, bound together using steel, concrete, and so forth; and is designed to resist lateral forces (sideways shaking).

Construction quality B includes good workmanship and mortar; has reinforcement; but is not designed to resist strong lateral forces.

Construction quality C includes ordinary workmanship and mortar; has no extreme weaknesses such as failing to tie in at comers; but is not designed or reinforced to resist lateral forces.

"Today's Standards" refers to restoring the buildings to conform to current earthquake-resistant design and construction practices. The incremental hazard to these buildings in relation to the recently constructed buildings is essentially negligible.

Thus, there are four upgrading alternatives for each of the four risk classes of the buildings. For the purposes of our analysis, Class III and Class IV are considered together since there is no significant difference between these two classes. Thus, we will evaluate 12 upgrading alternatives.

4.1.2 Attributes

The principal attributes in the decision problem are cost of upgrading, property damage, number of deaths, and number of injuries requiring hospitalization. The first two attributes can be combined to yield a cost measure, while the last two represent a composite safety measure. Occasionally we will use these aggregated attributes to depict cost/safety trade-offs.

4.1.3 Time Horizon

Any choice of a time horizon is somewhat arbitrary. We will consider 10 years as a planning horizon. This planning horizon is selected because the ordinance for the earthquake hazardous buildings developed by the city stipulates a 10-year period for a phased compliance with the code. A shorter time horizon would not reflect the earthquake damages accurately. A longer time horizon would require additional data on natural attrition of the buildings, and the possibility of more than one earthquake would have to be formally included in the analysis. It is believed that one major earthquake will result in the demolition of a large proportion of the old buildings (50%-90%) and stricter codes will be promulgated subsequent to such a disaster. Thus at the present time, a 10-year planning horizon is realistic and relevant for evaluating policy options. We do not include natural attrition among the masonry buildings in the next 10 years in our analysis.

4.1.4 Earthquake Scenarios

We examine four scenarios of an earthquake in the Los Angeles Basin. These are:

1. A IX MMI (7.5 Richter scale) earthquake along the Newport-Inglewood fault (Scenario 1).

2. A VIII MMI (6.5 Richter scale) earthquake along the Newport-Inglewood fault (Scenario 2).

3. A X MMI (8.3 Richter scale) earthquake along the San Andreas fault (Scenario 3).

4. No earthquake (Scenario 4).

Here MMI refers to Modified Mercalli Index, which is a measure of the intensity of an earthquake (Wood and Neumann 1931). Based on the U.S. Department of Commerce report (1973), the distribution of earthquake intensity in each council district is shown in Table 1.

Table 1: Distribution of Earthquake Intensity

	MMI Estimate		
	Scenario 1	Scenario 2	Scenario 3
East San Fernando Valley	IX	VIII	VIII
Hollywood Hills	VIII	VII	VI
SW San Fernando Valley	IX	VIII	VIII
Wilshire	IX	VIII	VII
West Los Angeles	IX	VIII	VII
Venice to Crenshaw	IX	VIII	VIII
Central San Fernando Valley	IX	VIII	VIII
South Central Los Angeles	IX	VIII	VIII
Central City	IX	VIII	VIII
SW Los Angeles	IX	VIII	VIII
Brentwood to Encino	IX	VIII	VI
NW San Fernando Valley	IX	VIII	VIII
Hollywood	IX	VIII	VII
East Los Angeles	VIII	VII	VII
Watts to San Pedro	IX	VIII	VI

In order to compute the probability of each of the four scenarios, we use the historical frequency of the occurrence of earthquakes along the two faults. Based on the historical record, a return period of 19 years or more is assumed for an MMI VIII earthquake in downtown Los Angeles. If the interarrival time between earthquakes is assumed to be exponentially distributed, then based on a return period of 19 years, the probability of a VIII MMI or greater intensity earthquake in $10\ years = 1 - e^{-10/19} = 0.41$. Based on the Federal Emergency Management Agency report (1980), the probability of Scenario 1 is 0.01 and the probability of Scenario 3 is between 0.2 and 0.5. We consulted seismologist Dr. Clarance Allen of the California Institute of Technology to seek his subjective probability. He believes Scenario 2 is likely to happen with a 0.1 probability. Therefore, Scenario 3 has 0.3 probability of occurrence (0.41 – 0.1 – 0.01). Dr. Allen's subjective probability for Scenario 3 was 0.25 and for Scenario 1 it was 0.01. We assume the probabilities of Scenarios 1 through 4 as 0.01, 0.1, 0.3, and 0.59, respectively.

It should be noted that these four scenarios are not the only possible scenarios for earthquakes in Los Angeles. These four scenarios are chosen simply because they are representative of the range of possible damaging earthquakes in the area. These four scenarios represent a collectively exhaustive set under the assumption that smaller earthquakes not included in Scenarios 1 through 3 would produce little damage.

Further, the probability estimates for various scenarios need refinement. One approach will be to treat these probabilities as uncertain variables and quantify second-order probabilities. A sensitivity analysis can also be conducted by varying the probabilities of alternative scenarios.

4.1.5 Interactive Decision Tree

The earthquake safety problem of the city of Los Angeles can be represented as an interactive decision tree shown in Figure 1. For each class of the buildings, the city could take any of the four actions shown in the decision tree. The owner may comply by acting consistently with the city's chosen action or may choose not to comply. For each class of building the owner also has four actions available to him. It is to be noted that under strict regulation, the owner is forced to choose the same action as the city dictates. But, for planning purposes it is worthwhile to explore what is the owner's most preferred choice and how this choice differs from that of the city.

Figure 1: Interactive Decision Tree

CLASS I

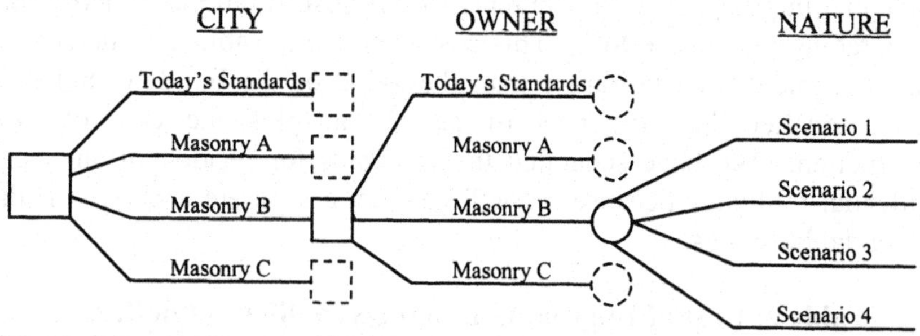

CITY
Today's Standards
Masonry A
Masonry B
Masonry C

OWNER
Today's Standards
Masonry A
Masonry B
Masonry C

NATURE
Scenario 1
Scenario 2
Scenario 3
Scenario 4

Consequences

	Cost of Upgrading (millions)	Property Damage (millions)	Deaths	Injuries
Scenario 1	$2.59	$2.5	30	150
Scenario 2	$2.59	$0.5	—	—
Scenario 3	$2.59	$0.4	—	—
Scenario 4	—	—	—	—

* Class II and Class III, IV decision trees as above.

The decision tree also depicts the four earthquake scenarios, only one of which will actually occur. The consequences on the four attributes for each action/scenario combination are represented at the end of each branch of the decision tree.

4.2 Estimation of Consequences

4.2.1 Cost of Upgrading

Costs of upgrading the buildings from their present Masonry C standard to Masonry B, Masonry A, and Today's Standard are $5/ft², $10/ft², and $20/ft², respectively. These costs are based on the engineering estimates of Wheeler and Gray (1980) who examined in detail some representative buildings. We took a sample of 61 residential and 60 non-residential buildings to estimate the area and the market value of these buildings. The average area of a residential building was found to be 22,471 ft² and for a non-residential building it was 8,783 ft². We also know the number of buildings in each class (Class I: 59 non-residential; Class II: 147 residential and 646 non-residential; Classes III and IV: 643 residential and 6,368 non-residential). The total cost of upgrading for each class of buildings is easily computed to be (22,471 *x* number of residential buildings + 8783 *x* number of non-residential buildings) *x* cost of upgrading per square foot. This cost is given in Table 2. The cost of upgrading may often be larger than the value of the building, but it is always smaller than the cost of demolition plus the cost of new construction. We have separated the estimates for residential and non-residential buildings because we will conduct a separate analysis for the residential buildings.

Table 2: Cost of Upgrading Buildings (millions of dollars)

	Today's Standard		Masonry A		Masonry B	
	Residential	Non-Residential	Residential	Non-Residential	Residential	Non-Residential
Class I	—	10.36	—	5.18	—	2.59
Class II	66.06	113.46	33.03	56.73	16.51	28.36
Class III & IV	289	1118.6	144.5	559.3	72.25	279.65

4.2.2 Property Damage

Property damage will vary for each upgrading alternative and for each scenario of the occurrence of an earthquake. Based on our sample of 121 buildings, the average value of a residential building exclusive of land is estimated to be $34,100 and that of a non-residential building is $101,520. These average building values seem quite low. There could be two reasons for this. One is that the records may indicate depreciated values rather than the market values. Second, the land may be assigned a disproportionately high value in the determination of the value of buildings. An appraisal of randomly selected buildings by real estate experts should be undertaken to determine the building values accurately. Property damage depends on the intensity of the earthquake and the standard of the building. Obviously, a lower standard and a higher intensity of earthquake would cause the greatest damage. In Table 3, the percentage of property value damaged under various intensities of an earthquake and for each of the four building standards is given. This table is based on the definition of MMI scale. A possible refinement of these damage estimates is to quantify uncertainty inherent in such estimates in the form of probability distributions. Professional judgments of engineers along with past data can be used to specify these probability distributions.

Table 3: Damage Factors (% of total value damaged)

Earthquake Intensity (MMI)	Masonry C	Masonry B	Masonry A	Today's Standard
VI	—	—	—	—
VII	10%	—	—	—
VIII	50%	10%	—	—
IX	90%	50%	10%	—

Since we know the distribution of the number of buildings in each council district as well as the intensity of earthquake under the four scenarios, the property damage is easily computed. Based on the opinion of a real estate expert, the value of the contents of the building is assumed to be 25% of the value of the property. The total amount of damages is thus 1.25 *x* property value damaged. These values are given in Table 4. (Los Angeles City Planning Department 1979, Table 23)

Table 4: Total Value of Property Damage (million dollars)

Class	Upgrading Alternative	Property Damage			
		Scenario 1	Scenario 2	Scenario 3	Scenario 4
I	TS	—	—	—	—
	A	0.5	—	—	—
	B	2.5	0.5	0.4	—
	C	4.7	2.5	2	—
II	TS	—	—	—	—
	A	6.4	—	—	—
	B	32	6.4	4.5	—
	C	60	32	25	—
III & IV	TS	—	—	—	—
	A	56	—	—	—
	B	277	56	40	—
	C	526	277	216	—

4.2.3 Number of Deaths

Deaths occur because of a collapse or a partial collapse of buildings. The number of deaths caused by an earthquake depends on the number of people exposed and the extent of the damage suffered by the buildings.

The number of people exposed to the falling debris depends on the time of the occurrence of an earthquake. From the occupancy data, it is seen that there is an average of approximately 125 people/building (*1 million people ÷ 8000 buildings*). We will assume that, on average, half of the occupants/building are exposed.

Table 3 provides an estimate of the percentage of building damage for varying magnitudes of an earthquake. Based on professional estimates, we assume that 90% or more building damage will cause 5% deaths, 50%-90% building damage will cause 2% deaths, and 30%-50% building damage will cause 0.2% deaths among the occupants of the buildings. Based on these assumptions, the number of deaths under each scenario of earthquake is given in Table 5. For a comparison, in Scenario 1, the Federal Emergency Management Association (1980) estimates 4,000-23,000 deaths; Soloman et al. (1977) estimate 13,000 deaths; the U.S. Department of Commerce Study (1973) estimates 4,000-20,000 deaths; and several other reported estimates in the *Los Angeles Times* news reports (1979; 1980; 1981) are around 12,000-15,000 deaths. In Scenario 3, the estimates from these sources range from 3,000 to 14,000 deaths. The estimates under Scenario 3 are lower in spite of it being a higher-intensity earthquake because of the larger distance between the fault and populated areas. While our estimates are within the range of other estimates, a considerable uncertainty exists in estimating exposure levels, deaths, and injuries. A desirable refinement of these estimates is to use experts' opinion to quantify this uncertainty. The experts' should, however, take into account the past data, the results of other studies, and their own professional judgment in providing their estimates.

Table 5: Number of Deaths and Injuries

	Upgr.	Scenario 1		Scenario 2		Scenario 3		Scenario 4	
Class	Alt.	Deaths	Injuries	Deaths	Injuries	Deaths	Injuries	Deaths	Injuries
I	TS	—	—	—	—	—	—	—	—
	A	—	—	—	—	—	—	—	—
	B	30	150	—	—	—	—	—	—
	C	145	725	30	150	20	100	—	—
II	TS	—	—	—	—	—	—	—	—
	A	5	20	—	—	—	—	—	—
	B	470	2,350	15	75	10	50	—	—
	C	2,195	10,975	47 0	2,350	3 60	1,800	—	—
III	TS	—	—	—	—	—	—	—	—
&	A	20	100	—	—	—	—	—	—
IV	B	2,040	10,200	80	400	65	325	—	—
	C	16,755	83,775	3,540	17,700	2,335	11,675	—	—

4.2.4 Number of Injuries

We have assumed the number of injuries requiring hospitalization to be five times the number of deaths. Other studies assume the number of injuries to be from four to six times the number of deaths. The estimates of the number of injuries are also given in Table 5.

4.3 Analysis

We will first conduct the analysis from society's viewpoint. Next, we will examine the problem from the viewpoint of the owners of the buildings who have to pay the costs for upgrading. We will then do the analysis using the assumption that the occupants of these buildings are informed of the risks involved and therefore they may pay a lower rent for higher risk buildings. Finally, we will consider the residential buildings separately.

4.3.1 Society's Viewpoint

In this analysis, we examine the total costs and total benefits disregarding to whom they accrue. In Table 6, the cost of rehabilitation, expected property damage, expected deaths, and expected injuries are given for each of the 12 policies. The expected values are computed by multiplying the outcome (such as number of deaths) under a given scenario with the scenario probability.

Table 6: Social Costs and Benefits

Class	Policy	Cost of Rehabilitation (millions of dollars)	Expected Property Damage (millions of dollars)	Expected Deaths	Expected Injuries
I	TS	10.36	—	—	—
	Masonry A	5.18	0.09	—	—
	Masonry B	2.59	0.58	0.3	1.5
	Masonry C	—	1.61	10.4	52
II	TS	179.5	—	—	—
	Masonry A	89.75	1.15	0.05	0.02 5
	Masonry B	44.9	7.45	9.2	46.0
	Masonry C	—	20.1	176.9	884.5
III	TS	1,407.6	—	—	—
&	Masonry A	703.8	10.08	0.2	1
IV	Masonry B	351.9	64.74	47.9	427.5
	Masonry C	—	175.26	1,222	6,110

The choice of a policy depends on the society's willingness to pay in order to reduce the number of deaths and injuries. The willingness to pay, however, depends on the risk that the individuals face. In Figure 2, a hypothetical curve illustrating the society's willingness to pay to prevent one expected death as a function of the probability of death is given. This curve is derived by using the range of "value of life" figures reported in the literature (Bailey 1980). This curve shows, for example, that if the individuals face a 5-in-1,000 chance of dying, then the society is willing to pay \$1 million to prevent 1 expected death. But, if the individuals face only a 5-in-10,000 chance of dying then the society will pay only \$500,000 to prevent 1 expected death.

Figure 2: Willingness-to-Pay Curve

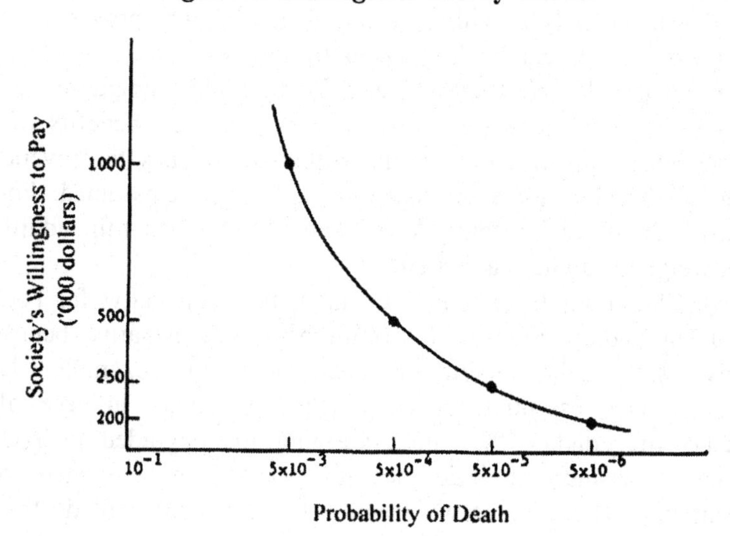

From Table 6 and using the approximate figures from the willingness-to-pay curve, the incremental costs and benefits are computed in Table 7. For precise computations, the willingness-to-pay curve should be analytically expressed and the willingness-to-pay in going from one policy to another (e.g., Masonry C to Masonry B) should be computed using integration. Our calculations are based on simple approximations.

Table 7: Incremental Social Costs and Benefits (millions of dollars)

		Cost	Benefits			
Class	Policy	Additional Cost of Upgrading	Property Damage	Value of Deaths	Value of Injuries	Benefit-Cost
I	A to TS	5.18	0.09	—	—	-5.09
	B to A	2.59	0.49	0.15	0.037	-1.9
	C to B	2.59	0.73	7.9	1.3	7.34
II	A to TS	89.75	1.15	—	—	-88.60
	B to A	44.9	6.30	4.57	1.15	-32.88
	C to B	44.9	12.65	125.7	22.1	115.55
III &	A to TS	703.8	10.08	0.02	0.025	-693.67
IV	B to A	351.9	54.66	23.85	10.68	-262.71
	C to B	351.9	110.52	880.5	152.75	791.87

A similar curve can be estimated for injuries. For simplicity, we will assume that the society is willing to pay $25,000/injury prevented. The following conclusions can be drawn from this analysis.

(1) For Class II and Class III and IV buildings, upgrading to the Masonry B standard deserves consideration. The net benefits of this upgrading remain positive even if the willingness to pay to prevent one expected death and injury is one-fourth of what we have assumed. For the other two alternatives (Masonry A and Today's Standards) the additional costs outweigh the additional benefits.

(2) For Class I buildings, though the net benefit is negative for Today's Standards upgrading policy, the magnitude of the negative benefit is relatively small. Other qualitative considerations, for example Class I buildings provide essential service to the community in case of an earthquake, may dictate that these buildings be upgraded to Today's Standards. Because of lack of data, we did not quantify these considerations. However, if we assume that the number of deaths and injuries prevented to the outside users of these facilities in case of an earthquake are equal to the deaths and injuries to the occupants of the buildings, then upgrading to Today's Standards is cost-effective.

(3) It should be noted that even though the net benefit in going from the Masonry C standard to the Masonry A standard is positive for Class II and Class III and IV buildings, it is not cost-effective to upgrade to Masonry A. This is because most of the benefits of upgrading are reaped in going to the Masonry B standard and the additional benefit of further improvement to Masonry A does not justify the additional cost.

4.3.2 Owner's Viewpoint

A typical owner faces the decision problem depicted in Figure 1 for Class II buildings. The city could take any of the four actions. The owner also has the four alternatives available. For every scenario, the alternative of not upgrading dominates the upgrading alternative because an owner is not liable for the deaths and injuries caused by a natural hazard. This finding is clearly supported by the owners' opposition to any ordinance that requires them to upgrade their buildings. Of course, if there is a sufficiently large penalty for noncompliance, then the owners can be forced to upgrade the buildings.

4.3.3 Empirical Estimation of Willingness to Pay for Safety

In this section, we will discuss two approaches that we used in estimating the willingness of the occupants of the hazardous buildings to pay for safety. In one approach, a questionnaire was used to elicit directly how much an occupant is willing to pay for a decrease in the probability of death and injury due to earthquake. In the other approach, the market-determined price of the earthquake-unsafe buildings was compared with the price for similar earthquake-safe buildings. The difference in the market price when adjusted for the quality of the building provided an estimate of how much of a premium the market is attaching for the safer buildings.

4.3.4 Direct Estimation of Willingness to Pay for Safety

In this approach, 12 residents were individually interviewed to determine their willingness to pay additional monthly rent if the buildings are strengthened. Each resident was asked background information on age, income, monthly rent, number of years in the building, and his general comments on the earthquake safety issue. To estimate willingness to pay the following question was asked:

> *Your building is known to be unsafe with regard to earthquakes. There are 10,000 people living in such buildings in your neighborhood. If nothing is done to strengthen the buildings, 10 of these residents are going to die in the next 10 years when the earthquake will strike. You or your family members could also become fatalities. Are you prepared to pay an increased rent to help strengthen the buildings so that 9 out of 10 residents would be saved? How much?* _____. *How much are you prepared to pay so that none of your 10,000 neighbors will get killed?* _____.

The above scenario estimates the willingness to pay for decreasing the probability of death from 1 in 1,000 to 1 in 10,000. Alternative scenarios were presented to estimate willingness to pay for different levels of decreases in the probability of death.

The results of this survey were as follows:

1. The willingness to pay in increased monthly rents varied from $0 to $25/month.

2. The willingness to pay did not depend on the initial probability of death. Respondents took the position that the amount paid is the same whether 1 of 10,000 people is killed or 10 or 100 of 10,000 people are killed.

3. Older residents (above 70) were willing to pay little or nothing for the improved safety.

This survey is illustrative as the sample size is clearly inadequate. Further, several alternative wordings of the questions to elicit willingness to pay should be tested. Our question may be biased for its deterministic tone.

4.3.5 Property Value Differential Approach for Willingness to Pay for Safety

In this approach, the market prices of pre-1933 buildings were compared with those of post-1933 buildings by choosing a sample of matched pairs. Both buildings in a pair were located adjacent or at least on the same block to control for neighborhood, environmental, and social service quality levels. Thus, the buildings in a pair differed on building quality, size, and seismic resistance. Fieldwork by project staff was undertaken to compile subjective ratings on the "differences in quality" between all matched pairs. The resulting "rating" variable became the independent variable (x) in the regression analysis. Size differences were taken care of by normalization: the dependent variable (y) of the regression became value per square foot of building space. The building value data came from the Los Angeles County Assessor's records.

The regression of a value-per-square-foot (y) on differences in quality rating (x) was carried out with a transformation on the dependent variable. The intercept term can be interpreted as property value differences that exist for the case of no difference in quality. Assuming an annualized yield of 10%, the average difference between earthquake-safe and earthquake-unsafe buildings was found to be \$0.41/ square foot. Thus, an 800-ft^2 apartment would rent for \$328/year (\$27/month) more if it were earthquake-safe, all other things held constant.

For greater reliability, more variables measuring dwelling-unit characteristics should be included in the analysis (Brookshire et al. 1982). If we accept the above results, it can be argued that the upgrading alternatives requiring more than \$4.10/ft^2 are cost-ineffective. Non-economic arguments such as willingness to pay of the unaffected residents for the safety of the residents of the hazardous buildings may justify upgrading to Masonry B (\$5/ft^2). The "non-economic" arguments have to be substantially stronger for the higher-cost upgrading schemes.

4.3.6 Analysis with Public Awareness of the Earthquake Hazard

In this analysis, we will assume freemarket conditions where there is no rent control by the city and the tenants are aware of the earthquake hazard of their buildings. Based on our empirical study, we further assume that a tenant is willing to pay \$25/month in additional rents for an

800-ft^2 apartment if it is upgraded (chances of a tenant's death in the next 10 years are reduced at least from 1 in 1,000 to 1 in 10,000). Thus, the increased value of rental is 37.5cents/ft^2/yr. The total area of the buildings is 83.946-million ft^2.

We also assume that the property damage can occur in any of the 10 years with equal probability and the cost of upgrading is incurred in the beginning of the planning horizon. The present value of the property damage and the increased rental income is calculated using a 10% discount rate. Present values of costs and benefits are summarized in Table 8.

Table 8: Present Values of Costs and Benefits (millions of dollars)

Class	Policy	Cost of Upgrading	Property Damage	Increased Rental
I	TS	10.36	—	1.5
	Masonry A	5.18	0.05	1.5
	Masonry B	2.59	0.35	1.5
	Masonry C	—	0.95	—
II	TS	179.5	—	21
	Masonry A	89.75	1	21
	Masonry B	44.90	5	21
	Masonry C	—	12	—
III & IV	TS	1,407.6	—	171
	Masonry A	703.8	6	171
	Masonry B	651.9	40	171
	Masonry C	—	108	—

From Table 8, it is easily seen that in aggregate all upgrading alternatives are unattractive in spite of the increased rents.

4.3.7 Residential Apartment Buildings

We now examine whether it is reasonable to have a separate policy for the residential buildings. Based on the Environmental Impact Report of the Los Angeles City Planning Department (1979), there are 137,000 apartment dwellers that live in 45,622 earthquake-unsafe units. We assume that on an average, two-thirds of the residents are exposed to the earthquake. This exposure estimate may be high for a normal population, but is reasonable in this case because a majority of the residents of these buildings are old and retired. Based on the exposure, we can now calculate the expected deaths and injuries as discussed earlier. We already provided the cost of upgrading for residential buildings in Table 2. Since we know the distribution of these buildings among council districts, the expected property damage can be calculated. The costs and benefits of the residential buildings are given in Table 9.

Table 9: Costs and Benefits of Residential Buildings

Policy	Cost of Upgrading (millions of dollars)	Expected Property Damage (millions of dollars)	Expected Deaths	Expected Injuries
TS	355.06	—	—	—
Masonry A	177.53	3.28	0.04	0.2
Masonry B	88.76	21.1	10.7	53.5
Masonry C	—	57.12	262.1	1,310.5

It is seen from this table that if the society is willing to pay $200,000/life saved and $10,000/injury prevented then upgrading to Masonry B is cost-effective (*benefits = [(57.12-21.1)+0.2(262.1-10.7) +0.01(1310.5-53.5)-88.76=10.11]*). Further, the additional cost of upgrading, net of property damage, for upgrading to Masonry B is $52.74 million (*88.76 -(57.12 - 21.1)*). This cost is easily recouped in 10 years even if for each dwelling unit the rents are raised by $20/month. The break-even rental increase at a discount rate of 10% for 10 years is approximately $16/month. In our survey, the residents were willing to pay this amount for increased safety.

5. RECOMMENDATIONS

Our study shows that the risks to the occupants of the unreinforced masonry buildings are significant. If no upgrading of these buildings is undertaken, an individual occupant faces approximately 5-in-1,000 chance of death, and 25-in-1,000 chance of serious injury due to an earthquake in the next 10 years. This risk is about 10 times the risk due to fire and flames and about 40 times the risk due to electricity current in the home during the same time period.

Our estimated total cost of upgrading these buildings to Masonry B standard is approximately $400 million. Of course an upgrading of these buildings will result in lower property damage to the owners of these buildings ($125 million savings), but this gain clearly does not offset the costs involved. A policy that does not account for the owners' interests has a low likelihood of success. Besides, the cost of implementing a policy that disregards the owners' interest would be tremendous. This is because the unwilling owners find all sorts of ways (legal, political) for not complying with the policy.

Past experience suggests that the owners have ignored the upgrading of the buildings because of the high cost of rehabilitation. There is also evidence that the city has been unable to enforce seismic design regulations because of financial problems and trained-manpower shortages. Therefore, a seismic safety policy should provide an incentive for the owners to cooperate. Keeping in view the interests of both the owners and the occupants of the buildings, we provide the following recommendations:

(1) Class I buildings constitute essential buildings such as schools, hospitals, fire stations, and so forth. These buildings should be upgraded to Today's Standards. A negative net benefit of upgrading reported in our analysis does not include the benefits to the general public due to uninterrupted operation of these emergency facilities in the event of an earthquake.

(2) Residential buildings should be upgraded to Masonry B standards. The net benefit of this policy is positive if an individual occupant is willing to pay $16/month for the reduced risk.

We feel that approximately $10/month/dwelling increase in rent is a fair cost sharing by the owners and the tenants. This is because the owner receives other benefits by upgrading, for example tax advantages, an increase in the life of the building, increased property value, protection against lawsuits, insurance benefits, and so on, that were not included in our calculations. The city should also ensure that adequate financing through conventional channels is made available to the owners for undertaking the upgrading.

We do not recommend that the city should simply post signs to make the residents aware of the hazard on the belief that the market mechanism will determine the optimal action. This is because for an average resident it is relatively difficult to assess the risks involved. Besides, because of the housing shortage in Los Angeles, in the short run, the residents may not have a real choice of paying a higher rent for a safer building. An ordinance based on a cost-sharing scheme between the tenants and the owners would reduce the resistance of the owners to upgrading. Such a scheme would therefore be beneficial to both the owners and the tenants.

(3) Buildings other than Class I and residentials should not be regulated. For these buildings, we recommend that occupants be made aware of the hazard. The final course of action should be allowed to be decided by the market mechanism. A scheme to inform the public about the seismic hazard of a building has been opposed by the owners of the buildings. It is our belief that the risks involved are substantial and

therefore it is the responsibility of the city to inform the public about the risks involved. We conjecture that some owners will decide to upgrade the buildings to avoid adverse public reaction and pressure from the occupants once earthquake hazard information is made public.

The city ordinance requires all buildings be upgraded to specified design standards that in our terminology amounts to an upgrading to somewhere between Masonry A and Today's Standards. The owners are given two options. In Option I, they must meet the standards within 3 years from the date they are notified to upgrade the buildings. The actual notification date varies depending on the building classification. In Option 2, the owners could undertake a reduced upgrading, wall anchoring (that corresponds to somewhere between Masonry C and Masonry B standards), within a year of the notification. Once this reduced upgrading is undertaken an additional 3-9 years are permitted for full compliance.

It is not possible to compare the relative success of implementation of our recommendations with the provisions of the city ordinance. It can however be said that the ordinance provides little, if any, incentives to the owners. As reported in the *Los Angeles Times* (1981), the owners oppose the ordinance. The owners' lack of cooperation will undoubtedly make the enforcement tedious.

The city ordinance does not distinguish residential buildings from commercial/industrial buildings. Our recommendations would provide the adequate safety to the residents while allowing the market mechanism, public opinion, and occupant/owner negotiations to determine the acceptable course of action for the non-residential buildings. One possible result may be that some buildings will be upgraded while some others will be put to alternate use with low people exposure, for example, warehouses, and so forth.

Finally, our recommendations are based on an analysis of the costs and benefits of each alternative. It is to be admitted that all costs and all benefits were not quantified in the formal analysis. Nevertheless, the results of a social decision analysis could be quite useful in formulating a policy.

The issue can be raised that a local government often lacks resources to conduct a detailed social decision analysis. It may therefore be more appropriate for a state or a federal-level agency to conduct an analysis as we have proposed. This issue has merit. We recommend that an extensive effort must only be undertaken for problems having significant impacts on the constituents, and where choice of a policy is not clear. In many situations, a quick and aggregate analysis along the lines of the approach discussed here may reveal a dominating policy. In some situations, available data can be used to establish whether a problem needs urgent action or simply occasional monitoring. In conclusion, we recommend that even if a local government lacks resources to conduct social decision analysis, the steps of our approach provide a guideline for a discussion of various aspects of the problem.

6. CONCLUSIONS

In this paper, we provided a social decision analysis for the seismic safety problem faced by the city of Los Angeles with regard to its old masonry buildings. Costs and benefits of alternative policies were compared for the society as a whole and for the affected constituents. Both regulatory and freemarket alternatives were evaluated. The trade-offs between additional cost and safety were determined by using a direct willingness-to-pay approach and an economic approach based on property-value differential.

It is our belief that the solutions for a societal problem that affects a large number of citizens should not be left solely to the intuitive determination of bureaucrats and politicians. A formal analysis, while not able to resolve the complex value questions completely, goes a long way in pointing out socially desirable policies. A social decision analysis integrates scientific facts and the value trade-offs of the impacted constituents. Thus, it provides a useful insight into various dimensions of the decision problem and hopefully has the potential to aid decision-makers, as well as various effected constituents, in the process of reaching an acceptable solution.

REFERENCES

Bailey, Martin J. 1980. *Reducing Risks to Life: Measurement of the Benefits.* Washington, D.C.: American Enterprise Institute.

Bolt, Bruce A. 1978. *Earthquakes: A Primer.* San Francisco, CA: W. H. Freeman.

Bolt, Bruce A. 1978. "Earthquakes hazards." *EOS: Transactions, American Geophysical Union* 59(11).

Brookshire, D.S., M. A. Thayer, M. A. Schulze, D. William, and R. C. D'Arge. 1982. "Valuing Public Goods: A Comparison of Survey and Hedonic Approaches." *The American Economic Review* 72(l): 165-177.

Federal Emergency Management Agency. 1980. "An Assessment of the Consequences and Preparations for a Catastrophic California Earthquake: Findings and Action Taken."

Green, Paul E. and V. Srinivasan. 1978. "Conjoint Analysis in Consumer Research: Issues and Outlook." *Journal of Consumer Research* 5(September): 103-123.

Keeney, Ralph L. 1981. "Analysis of Preference Dependencies among Objectives." *Operations Research* 29(6).

Keeney, Ralph L. 1982. "Decision Analysis: An Overview." *Operations Research* 30: 803-838.

Keeney, Ralph L. and Howard Raiffa. 1976. *Decisions with Multiple Objectives*. New York, NY: Wiley.
Kerr, Richard A. "California's Shaking Next Time." *Science* 215(January): 385-387.
Los Angeles Times. November 25, 1979.
Los Angeles Times. July 27, 1980.
Los Angeles Times. January 5, 1981.
Los Angeles City Planning Department. 1979. *Draft Environmental Impact Report*. EIR No. 583-78 CW (September).
Raiffa, Howard. 1968. *Decision Analysis*. Reading, MA: Addison Wesley.
Sarin, Rakesh K. 1982. "Risk Management Policy for Earthquake Hazard Reduction," Report prepared under NSF Grant 79-10804, UCLA-ENG-8244.
Scott, S. 1979. *Policies for Seismic Safety: Elements of a State Governmental Program*. University of California, Berkeley: Institute of Governmental Studies.
Slovic, P., B. Fischoff, and S. Lichtenstein. 1980. "Facts and Fears: Understanding Perceived Risk." In *Societal Risk Assessment: How Safe is Safe Enough?*, edited by R. Schwing and W. A. Albers. New York, NY: Plenum.
Solomon, K.A., D. Okrent, and M. Rubin. 1977. "Earthquake Ordinances for the City of Los Angeles, California: A Brief Case Study." Report UCLA-Eng.-7765 (October).
Spetzler, C.S. and Christina Staël von Holstein. 1975. "Probability Encoding in Decision Analysis." *Management Science* 22: 340-358.
U.S. Department of Commerce. 1973. "A Study of Earthquake Losses in the Los Angeles California Area." Stock No. 0319-00026.
Wheeler and Gray. 1980. "Cost Study Report for Structural Strengthening Using Proposed Division 68 Standards." Prepared by consulting engineers under a contract awarded by the Department of Building and Safety, City of Los Angeles.
Wood, H.O. and F. Neumann. 1931. "Modified Mercalli Intensity Scale of 1931." *Bulletin of the Seismological Society of America* 21: 277-283.

APPENDIX

A.1 An Update on the Status of Unreinforced Masonry Buildings in Los Angeles

The city of Los Angeles enacted the Earthquake Hazard Reduction Ordinance, commonly known as Division 88 (the numerical section of the city code) in 1981. The ordinance required that the owners of the unreinforced masonry buildings built prior to October 6, 1933 make structural modifications over a stipulated time period. The time period to bring a building to the code varied depending on whether the building is deemed essential (school, hospital, etc.) and on a combination of occupancy load, historical importance, and structural integrity.

The ordinance applied to all unreinforced masonry buildings in Los Angeles (approximately 8,000 building). By 1996, one-third of the buildings were vacated or demolished. It is likely that in most of these cases the economics of seismic retrofitting was unfavorable to the owners. That is, the owners found that the initial expense of upgrading was not justified by possible rent increases and resale value. In some cases, the demolition was profitable as the cleared site could be put to alternate use. Of the remaining two-thirds of the buildings, 95% were in compliance with the ordinance by 1996. This is indeed a remarkable progress. Further, retrofitting tended to raise resale values by 37% and thus, over time the owners more than recouped the costs imposed on them by the ordinance. The owners also gained through increased rents. On average, a 20% rental increase was granted as the units were rent controlled. In retrospect, it is clear the owners' fear about the high cost of retrofitting with low potential future benefit was unjustified. Though the most unprofitable units were vacated or demolished, the units that were upgraded in accordance with the ordinance reduced risks to the occupants and provided profitable returns to the owners.

On January 17, 1994, a 6.8 magnitude earthquake (on the Richter scale) centered at Northridge struck Los Angeles. This unfortunate event provided a test case for the performance of the retrofitted buildings. It was observed that unretrofitted buildings generally had more extensive damage than retrofitted buildings. In fact, no properly retrofitted, unreinforced masonry building suffered significant structural damage. The end result of the ordinance and its enforcement has been that unreinforced masonry buildings no longer pose as significant a risk as they did prior to the enactment of the ordinance to the residents of Los Angeles.

8

Improving the Safety of Urban Areas through "Local Seismic Cultures":

From the Umbria 1997 Earthquake to Co-operation in the Mediterranean

Ferruccio Ferrigni
University of Naples "Federico II"
International Institute Stop Disasters

1. INTRODUCTION

1.1 Risks and Local Risk Culture

It almost appears trite to state that local building techniques in known areas at risk nearly always present some protective features. Yet not all risks lead to "anti-risk" techniques taking firm root. For example, flooding and avalanches, which usually affect the same areas, have conditioned the choice of sites for habitation purposes but have had little effect on the building techniques employed. Also, forest fires or landslides, which occur without any clearly definable regularity or characteristics, do not seem to produce risk awareness in people and consequently do not give rise to the development of specific techniques or behaviors in the communities affected.

In seismic risk areas, however, the regularity of the event can lead to building techniques and human behavior with a clearly identifiable protection function gradually taking root. Often these techniques can be seen in religious building works, public buildings, superior class housing or large-scale civil engineering works (bridges, aqueducts).

The earthquake-resistant properties of traditional Chinese architecture (Hu Shiping, 1991) [Fig. 1], of ancient Japanese pagodas (Tanabashi, 1960) [Fig. 2], have all been widely documented to date. It goes without saying

P.R. Kleindorfer and M.R. Sertel (eds.), Mitigation and Financing of Seismic Risks, 159–175.

that monuments and large-scale works were constructed using anti-seismic techniques. They were built to last. But the technical know-how employed, especially in the past, was an integral part of the local heritage. Therefore there is no reason to believe that the techniques adopted for "important" buildings were not known to whoever constructed ordinary buildings and that certain peculiar technical features are also not present in the latter.

Figure 1: The energy-dissipating system of the bracket set in a Chinese XVI century pagoda *(from Shiping Hu, modified)*

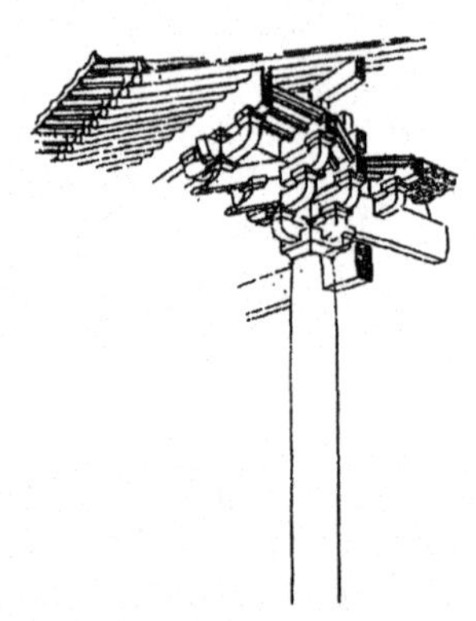

Figure 2: The same system of fig 1 in a Japanese pagoda of XV century (from B. Walker, modified)

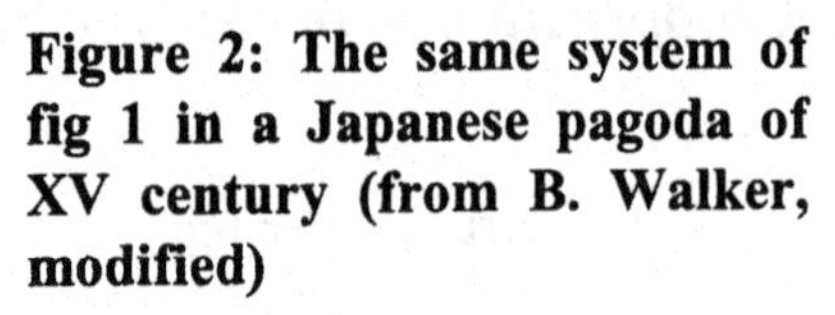

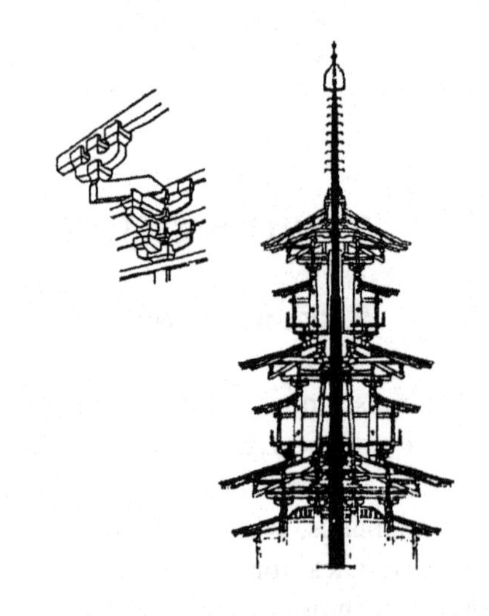

In fact, historical buildings – and, more generally, the "vernacular architecture" of the earthquake-prone regions - have been able to withstand earthquakes thanks to specific anti-seismic techniques used to construct them. A rich body of literature even provides evidence that traditional anti-seismic techniques dating back thousands of years (Shiping Hu, 1991; Touliatos, 1992) [Fig. 3] have survived to this day with few changes and have generated widespread "standard" practices and mechanical models clearly recognizable today (Giuffrè, 1993; Cardona, 2001).

Figure 3: In Lefkas island (Greece) a double structure saves human lives and enable damages to be rapidly repaired (from Touliatos, 1993)

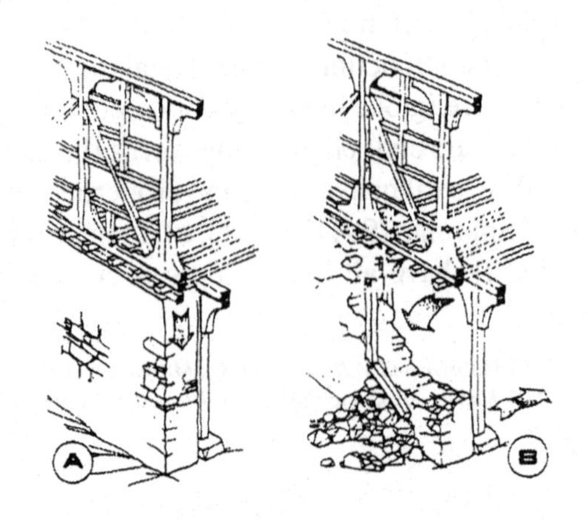

The ability of a building to withstand seismic shockwaves stems not only from the use of specific construction features. It also depends on the repairs and/or modifications made to the building as well as the choice of location for the actual village (which is almost always the safest in the area). A "Local Seismic Culture" (LSC) tends to develop in areas, which are regularly struck by earthquakes. We can define the LSC as a *combination of the knowledge of traditional anti-seismic techniques and appropriate behavior deriving from and consistent with this knowledge.*

Many earthquakes clearly illustrated that the majority of damage caused to towns and cities, in addition to almost all the human casualties, often resulted not just from the fact that ancient "rules" were not adhered to, but also that current anti-seismic regulations were simply not observed. This was because the former were considered obsolete or, more often than not, because they were only common knowledge to experts, and because the latter were looked upon as techniques which only increase costs and do not provide any real benefits. In short, local cultures have lost their knowledge of tried and tested local techniques and, more importantly, of how effective they are against the specific, local "seismic style." Thus modern regulations have devalued traditional techniques without imposing the generalized use of "new rules." Although they were to take the place of time-honored technical know-how developed in order to reduce risk, they have thus far failed to give birth to a fresh "culture."

The mosque of Golcuk, surviving in a totally destroyed environment, is a paradigm of difference between the modern "seismic culture" and the old one, [Fig. 4]. In shantytowns there is not even the slightest trace of the former consciousness of risk, and anti-seismic regulations have unfortunately not become a new culture.

Figure 4: The old and the new "Local Seismic Culture."

2. SEISMIC PREVENTION POLICIES, DECISION MAKERS' BEHAVIOUR AND BUILT-UP VULNERABILITY

2.1 The Umbria's lesson

The fact that regulations not in keeping with tradition may increase the vulnerability of the historic architectural heritage was recently confirmed by the earthquake which struck in Umbria (Italy) between 26 Sept. 1997 and 14 Oct. 1997. There the walls of historical houses are sack-like structures because the interstices between their outer, more or less regular stone-faces are filled with mortar and small stones. This "poor" and hardly correct technique (the two faces of a wall are barely connected with each other, see Fig. 5) had nonetheless preserved the buildings intact. After the 1984 earthquake many of them were "reinforced" by replacing the original wooden floors and roofs with reinforced concrete structures. But older buildings in Umbria are listed as historical preservation sites, and they cannot be outwardly altered. So, the engineers and government bodies in charge of the actions worked out a brilliant solution that would enable them to comply both with anti-seismic building codes and the regulations governing the

Figure 5: The historical "current" masonry of Umbria (Italy) is very fragile, although the region is highly seismic

Figure 6: To reconcile static and aesthetics, floors and collars beams lean only the interior part of the walls

protection of buildings categorized as historical monuments. Roofs, floors and platband were fixed in such a way as to rest on the inner wall face only [Fig. 6]. Consequently the 1997 earthquake impacted structures, which had already been thoroughly altered. The original wooden floors were lightweight structures loosely connected to each other. They absorbed a

Figure7: The different dynamic behavior of wood/concrete floors under the seismic shock

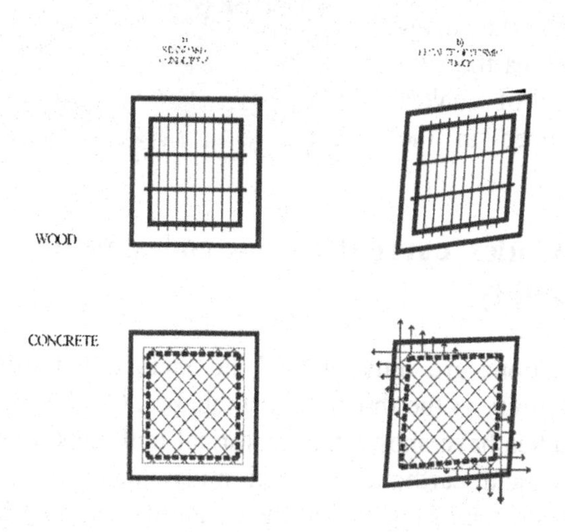

limited amount of seismic energy and they dissipated the greater part of it by deforming (Fig. 7/a). Therefore, they passed on negligible impact to the outer supporting walls and could even second the deformations caused in them. Conversely, the heavier and more rigid floors built in compliance with current regulations absorb most of the seismic energy without dissipating it. Unable to deform, they pass on the energy to the walls and provide a "punching" effect by pressing into the walls from the inside to the outside [Fig. 7/b]. In the event of a low-magnitude earthquake it is only the outer faces of the walls that fall [Fig. 8], but often the walls collapse altogether because they are unable to withstand the additional stress. Floors and roofs remain intact, but ...... they land just a few floors below [Fig 9].

Figure 8: The results of the "reinforcements"

In conclusion, all "reinforcement" actions were performed in compliance with current anti-seismic building codes. The new structures stood up well to the fresh quakes. However, the historical built-up heritage was irredeemably destroyed, although the magnitude of the 1997 earthquake did not exceed that of quakes previously experienced in the area.

Fig 9: Tested seismic code: the new floor (reinforced) is intact, the ancient building .

2.2 Ancient rules, current seismic codes, future vulnerability

Technical regulations are only one of the factors that influence the response of a community to the threat of earthquakes. On the one hand, the findings of past studies have shown that historical buildings in earthquake-prone regions withstand earthquake shock much better than is usually assumed. On the other, they suggest that their stability can only be properly assessed and further improved through the use of particular techniques. That the latter must be revived with the active participation of the communities concerned and made mandatory within the framework of town-planning instruments and anti-seismic building codes. That the modes of behavior of decision-makers in this field play a far from negligible part in magnifying/reducing the vulnerability of historical buildings (and modern constructions erected using similar technologies).

Today in many seismically active countries, once very rich in local seismic-proof techniques, a combination of circumstances contributes towards making ancient ordinary buildings even more vulnerable then they would be elsewhere and the new buildings much more vulnerable than the ancient ones.

In urban areas, two equally dangerous phenomena occur. In historical town centers the original constructions, which were almost always built according to anti-seismic criteria, are often in total disrepair. Because they are inhabited by low-rent tenants or because in the process of their gradual decay they were repartitioned and modified without paying any heed to their original characteristics.

In the suburbs, newcomers tend to settle on lower quality sites, which almost always prove the worst kind for building on. Houses are erected using rough and ready techniques by people from other areas who, more often than not, have no specific knowledge of any construction skills. Anti-seismic regulations - when they actually exist - almost never take advantage of or even refer to traditional anti-seismic techniques. Indeed, often they even make it impossible to apply them (this is a constraint, which also characterizes anti-seismic regulations in developed countries with a long-standing history of earthquakes such as Italy, Greece and Turkey). When houses are built with a total disregard for planning regulations it is highly unlikely that the owners pay any heed to anti-seismic regulations.

It is small wonder that little good has been achieved in this context by the repeated campaigns aimed at stirring the consciousness and sensibility of technicians and property owners to make them respect tried and tested construction rules. Therefore, any actions designed to recover "Local Seismic Cultures" will thus contribute to reducing the vulnerability of the system. A further important insight is that the modes of behavior are influenced by the policies enforced to address seismic risk and not by public information campaigns.

2.3 The (potentially) disastrous impact of the protecting laws

In earthquake-prone regions historical buildings constructed with traditional techniques are – of course - "earthquake-resistant." Provided they are properly altered and, above all, if they are constantly maintained, they are likely to survive an earthquake within the locally expected range of magnitude [Fig. 10]. On the other hand, although it is well known that maintenance and/or preventive reinforcement costs are far lower than reconstruction expenses, allocation for maintenance programs is usually scant while reconstruction aid is generally made available without great difficulty and in fairly sizeable amounts.

Thus the question we have to answer is why "anti-seismic prevention policies" are not an integral component of our current "culture" even though they are considered the most effective means of reducing the vulnerability of the historical built-up heritage. To a large extent this situation can be traced back to the fact that in many earthquake-prone countries preventive reinforcement is left to individual owners while reconstruction is funded by the community as a whole. No wonder, therefore, that the system adopts a wait-and-see policy instead of effectively providing against earthquake damage.

Figure 10: Usually in earthquake prone regions the vernacular architecture embodies some seismic-proof techniques. The original strength of the buildings is higher than the average. But if the Local Seismic Culture has been lost, inappropriate modifications, lack of maintenance and earthquakes reduce progressively the global resistance. Under the next earthquake the building collapses. (From Touliatos, 1993 modified)

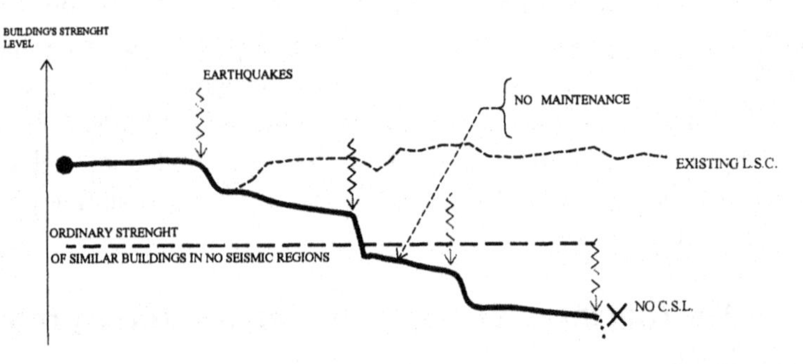

A comparison of two Italian laws, which reflect two different political approaches (preventive maintenance or reconstruction), confirm this remark. The first law, providing subsidies to restore existing buildings, was promulgated in 1978 (Act 457/78). The aim of this law was to redevelop the urban decay areas but, of course, the law may also be used for preventive anti-seismic reinforcement interventions. The second law (Act 219/81) was promulgated after the Irpinia 1980 earthquake and provided reconstruction aid. Therefore, we can consider it specifically oriented to post-event reinforcement[1].

An immediate remark stands out. The financial resources provided by the first law (457/78) are limited, but have still not been fully used. The resources of the second law (219/81) were enormous, but ten years after the earthquake, these have still not covered the demands of owners and local politicians. Why did this happen? Both are subsidizing laws, without technical instructions. So there is no problem of technical constraints. A preliminary analysis of the specific elements characterizing the two laws, however, shows a key difference between their disbursement procedures. According to 219/81, the Government gives the funds directly to the mayor, who then gives them to owners. The procedure established by Act 457/78 is different. The Government gives the funds to the Regional Minister and

---

[1] The 219/81 act concerns only the Irpinia earthquake, but laws concerning other earthquakes are very similar.

he gives them to the owners. Comparing the two "disbursement paths" with the flux of electoral support (Fig. 11), we see that the 219/81 procedure is centered on the mayor, a key figure in channelling the support of local electors to the national ministers. The procedures established by the Act 457/78, instead, don't concern the mayor. And the Regional Minister is too far from owners. Consequently, Law 219/81 drives mayors to obtain funds, while they are not interested in doing so under the provisions of 457/78. This difference helps to explain the low utilization of 457/78 funds for preventive seismic reinforcements.

Fig. 11: The comparison between the flux of the subventions and the flux of electoral support can explain why the reconstruction funds (act 219/81) are completely exhausted, while the maintenance funds (act 457/78) are not nearly fully used.

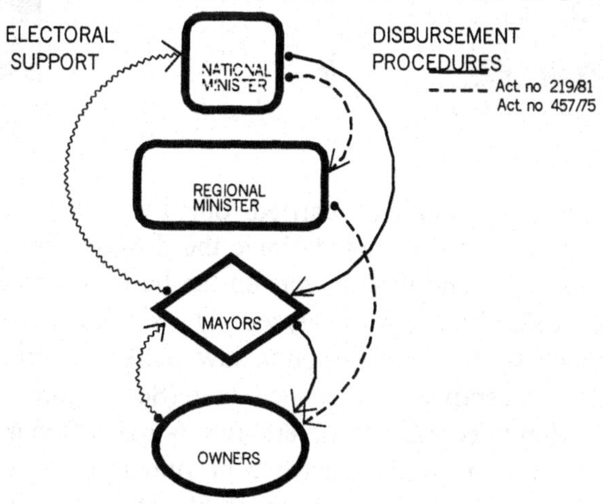

Alongside the analysis of traditional anti-seismic techniques, the "LSC research group" attached to International Institute Stop Disasters in Naples (IISD) and the European University Center for Cultural Heritage in Ravello (EUCCH) also addressed the issue of how policies for dealing with seismic risk may increase or decrease the vulnerability of historical built-up heritage. The effects on vulnerability caused by the above-mentioned laws have been investigated. First the *interests* of the *agents* (decision-makers) concerned were analyzed. Subsequently - using an approach, which combined multi-criteria and multi-purpose analyses - the "*overall satisfaction*" of various stakeholders with these laws was obtained by a survey,[2] and the benefits for the system as a whole, were simultaneously assessed.[3] The results of this comparative assessment via survey showed that compared with a situation in which no policy for dealing with seismic risk is enforced the post-earthquake reconstruction law increases the system's total "satisfaction" by 85 points. This means that the increase in satisfaction induced by this law (219/81) is twelve times that induced by the maintenance law (457/78, +7 points) (see Figures 12/a & b).

Figure 12: The "satisfaction" of all decision makers related to different laws favoring reconstruction versus prevention policies

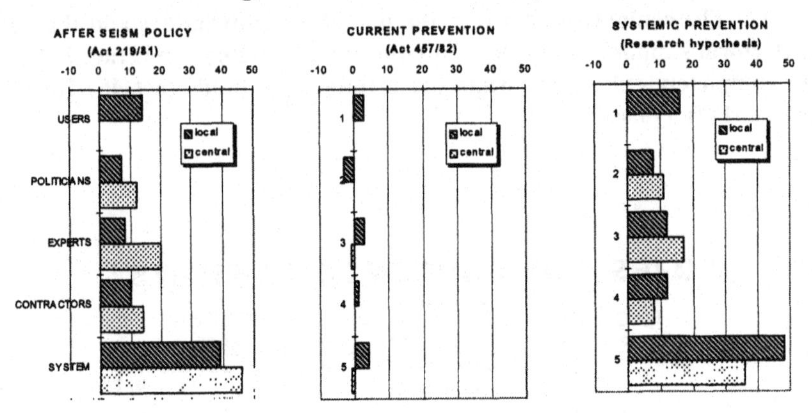

At this point, by analysis carried out by some case-studies and supplemented by the surveys noted above the different modes of behavior induced by these laws and their effects on the built-up heritage in terms of increasing its vulnerability was analyzed. The simulation showed that the behavior induced by the reconstruction law increases vulnerability much more than the preventive maintenance law (See Figure 13). The latter causes a fairly slight decrease in vulnerability (- 1 point) in more prosperous systems while a non-negligible increase in vulnerability results in poorer systems (+ 8). In general, these survey results show that the attitudes that parties with similar interests assume under the pressure of both these laws

---

[2] As Law 457/78 is valid everywhere in Italy, while Law 219/81 can be used only in regions stricken by the Irpinia 1980 earthquake, the comparisons reported apply only to the Irpinia region.

[3] The measures were conventional scores without any absolute value, but the huge gap between them may account for the fact that earthquake-prone countries that expend large amounts on reconstruction are slow to enforce seismic prevention policies providing for the permanent maintenance of the built-up heritage. The survey instrument and details of the attributes measured are available for the author.

causes a much greater increase in vulnerability (from 1.5-fold to 10-fold) in poorer systems than in prosperous ones.

This analysis sheds light on a few key factors. Firstly, our current "seismic culture" favors post-disaster intervention policies over prevention not because of insufficient foresight on the part of decision-makers, including politicians, but because the former results in greater benefits for the system as a whole. Secondly, these policies add to the vulnerability of the historical built-up heritage despite increasing aggregate expenditures on historic buildings. Thirdly, even maintenance aid policies are sure to produce this kind of impact on less prosperous systems if they are defined without due regard to the "interests" of the parties concerned.

In short, the research provides objective evidence in support of the assumption that the vulnerability of the historic architectural heritage, far from being merely a function of technical norms, numerical parameters and resources available, is largely determined by the policies which are enforced for dealing with seismic risk. At any rate, the most innovative and stimulating insight was the primary role played not so much by the amount of resources made available, as by the *disbursement procedures enforced.* They condition the attitudes of the parties involved and exert a direct impact on the vulnerability of the historic architectural heritage.

Figure 13: The vulnerability's variation depending on actor's behaviors induced by different political approaches

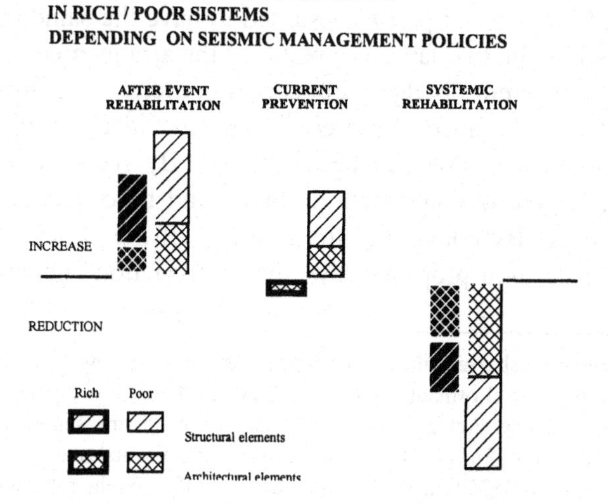

The comparative analysis of two laws in force in Italy has enabled us to *measure* their respective effects, which can be summed up as follows:

- a policy with special focus on post-earthquake repair/reconstruction aid is significantly more "satisfactory" to the parties concerned than a policy centered on preventive reinforcement actions;
- politicians acting on a national level always draw more substantial benefits from post-earthquake reconstruction policies than local politicians;[4]
- major construction firms and well-known technical experts will always reap greater advantages than small-scale local firms or local technical experts;
- the funds made available within the framework of such policies may either be earmarked in full or remain partially unused depending on the procedures with which the relevant subsidies are disbursed, i.e. from central government to local government and finally to property owners, and on whether the flows of electoral support - from electors to local or national politicians - are conveyed to those responsible for the relevant decisions.

The "disastrous impact" of the law providing for government-funded post-earthquake reconstruction and the obvious shortcomings of the law enforcing maintenance subsidies suggests using a different approach to mitigating the effects of seismic risk. An approach founded on the awareness of the decisive role that the LSC can plan and the behaviors induced in the parties concerned may play in securing successful prevention actions.

Setting out from a set of parties assumed to have the same interests, the effects produced by the two laws as a result of the arrangements as provided for in them (different budget allocation criteria, aid disbursement procedures, etc.) have been explored. Then, the ideal provisions for a "systemic maintenance" law has been defined. By recourse to the same procedures the degree of "satisfaction" this could be expected to produce has been estimated. By comparing the interests of the various actors with the effects induced by a proposed law, the "satisfaction" generated within

[4] The mayor-centered disbursement path (219/81) is more satisfying than the 457/78 both for central and local level officials, but the satisfaction increase is bigger for the central level than the local one. Actually, while Law 219/81 raises the satisfaction of central politicians by 12 points (from – 6 to + 6, see Tab 01) and mayors' satisfaction by 7 points (from – 2 to + 5), Law 457/78 make does not have any effect on the satisfaction of central politicians (it remains constant at – 6), but reduces the satisfaction of mayors (from – 2 to – 5).

the system (+84 points) was roughly on a par with that of the reconstruction law (+85 points), but far superior to that of the maintenance law currently in force (+ 7 points) (Figure 12/c). The same qualitative assessment procedures, based on case studies and survey methods, were used to analyze the variation in vulnerability caused by the two existing laws. These results show that the modes of behavior induced by the proposed law would greatly reduce the vulnerability of historical buildings. More importantly, this effect would be much greater in less well-off systems (- 18 points) than in fairly prosperous ones (- 11 points) (see Figure 13/c).

3. THE INVOLVEMENT OF INSURANCE COMPANIES IN SEISMIC PREVENTION: POTENTIAL AND DANGERS

3.1 Seismic codes, knowledge of engineers, role of insurance companies

According to the proposed "systemic maintenance law," loss control is primarily the responsibility of insurance companies. Simulations show that this procedure can be effective to stimulate the owners to reinforce the buildings, but demands a change on seismic codes and averts the risk of increasing vulnerability and inducing other improper practices.

First point. Usually, the aim of a seismic code is to reduce the human losses. Consequently, the technical rules of loss control concern mainly the structural elements. But insurance companies are interested to reduce overall economic losses, and not just those resulting from a lack of structural integrity. Thus, the standard technical rules of seismic codes are not well aligned with the interests of insurance companies. On this point the experience of Colombia is very interesting. The companies showed little interest in the requirements of a first law related to the coverage of seismic risks by the insurance. Then, the law was modified to introduce rules for the protection of non-structural elements. Now, insurance companies are collaborating with government and universities to calculate the premiums and are implementing a large advertising campaign.

On the second point, it goes without saying that insurance premiums will be higher the more vulnerable a building is assumed to be. If the widespread, but potentially erroneous, assumption that historical buildings in earthquake regions are more vulnerable than modern ones is not disproved, two equally vicious circles might be created. On the one hand the higher premiums charged by insurance companies in respect of traditional buildings might discourage property owners from entering into insurance policies and/or maintaining buildings which they think are not likely to stand the test of time. On the other hand owners might be tempted to reinforce their property by recourse to modern techniques, which are not necessarily compatible with the historic architectural heritage. In either case the result would be increased vulnerability.

To reduce this "supplementary risk," two actions appear useful. First is to train engineers on the seismic behavior of traditional local seismic-proof technologies and on the most appropriate techniques to up-grade these. This action would be carried out not only in the universities but also, above all, in cooperation with experienced engineers. In other words, the retrofitting would concern both the buildings and the engineers. Second, it is important to review and synthesize traditional seismic mitigation techniques. Such techniques were passed on typically by word of mouth and apprenticeship. So today knowledge of the effectiveness of many traditional anti-seismic techniques has been lost. Mediterranean countries have mutual cultural roots but present a large variety of vernacular architectures, marked by different seismic-proof techniques, depending on local available resources and local "seismic style." This circumstance offers an extraordinary opportunity to re-discover the "transnational rules" of the "traditional local seismic codes" and, by this way, to recover (and to up-grade) the LSC. Such an initiative would improve the exchange of information on the Mediterranean traditional local seismic technologies and could be an important and useful contribution both to understanding and preserving our historical heritage and to reduce vulnerability of ancient buildings.

3.2 The Turkish case, in the frame of Mediterranean cooperation

The Turkish region is very rich in monuments and in seismic-proof vernacular architecture. This is a favorable condition to stimulate an action aimed at reducing vulnerability of ancient buildings by recovering the LSC, and perhaps also by involving the insurance companies. To interest them, it is important that seismic codes protect both structural and *non-structural elements.* To avoid the risk of increasing vulnerability by using inappropriate technology, owners would be encouraged to "reinforce" or maintain their houses using techniques similar to those adopted by their original builders, at least to the extent that these techniques could be verified as having a sound engineering basis. The insurance companies would also then discover that local traditional techniques actually could decrease vulnerability, if these are used critically. Such a policy would be more effective if it were enforced within the framework of carefully developed official "analysis/design protocols," aiming to identify and to up-grade the local, traditional seismic-proof technologies. To support this policy, a small proportion of periodical "exceptional" earthquake damage repair allocations would be enough to fund policies that could drastically reduce vulnerability. Moreover, the greater part of the architectural heritage in earthquake-prone regions pre-dates the widespread introduction of reinforced concrete. The traditional maintenance is a highly labor-intensive process. Therefore, a seismic prevention policy with specific focus on traditional maintenance techniques could not only cut repair costs, but could also provide overall benefits far superior to the investments made.

This discussion of the similarities of the Turkish and Italian situation highlights the importance of engaging in an on-going discussion of the principles, methods, and techniques that might promote effective seismic prevention policy based on recovering the LSC and on involvement of the "local communities" (owners, politicians, experts, insurance, etc.). Above all, this action should be linked to trans-Mediterranean programs, like the MEDA Programs and EUR-OPA Major Hazards, offering a rich panel of actions and Pilot-Projects on seismic risk mitigation. Or the *Co-operation Platform*, mentioned by the Final Document of EUROMED-SAFE '99 (Naples, 27-29 October), promoting "the co-operation aimed at the full recognition of the local existing technologies and approaches which have demonstrated to be often effective in mitigation local risks."

4. CONCLUSIONS

Under the circumstances prevailing today, the enforcement of policies specifically focused on seismic prevention through permanent and effective maintenance is no longer a myth. Important international organizations are implementing large-scale programs aimed to reduce the impact of disasters and prepare the local populations to cope with emergencies, as well as development aid schemes founded on the use of local resources and the revival of traditional cultures (human growth). In point of fact these policies are strongly biased in favor of poorer countries. On closer analysis they may also work toward an objective to which the governments of all nations are striving: combating the generalized surge in unemployment. This trend is arguably hastened by a marked tendency toward investment in financial activities in preference to the production of goods and services and by the adoption of less and less labor-intensive production processes.

In this process some of international organizations - for example UNDP, OMS, UNESCO, NATO - might encourage national governments i) to test methods for reviving traditional anti-seismic techniques and the relevant aid allocation and disbursement procedures in purposely developed pilot schemes and ii) to enforce anti-seismic prevention schemes within their development aid programs (for instance by making it mandatory for such schemes to be developed with specific focus on the revival of traditional local anti-seismic building techniques). To make more effective the prevention policy, the national governments might sponsor and fund research/action projects involving local communities alongside with researchers. To help experienced engineers, whose formal education is well behind them, to design correctly, technical manuals and guidebooks on local traditional seismic-proof techniques (and the most appropriate up-grade techniques) would be produced. The technical support to engineers and the participatory action-researches could help the communities themselves to gain greater awareness of their cultural identity. Consequently, as the traditional seismic-proof rules characterizing the vernacular architecture can be used to retrofit (or to build) the non-engineered built-up, recovering the LSC can facilitate to reduce the vulnerability of the system.

The benefits of such an initiative would not only be to reduce the vulnerability of the system, but also to initiate a process toward sustainable growth founded on the use of local resources and to shift focus toward public and private investment in highly labor-intensive activities.

In conclusion, the mainstays of such a policy could be:

- TECHNICAL BUILDING REGULATIONS with a lesser focus on mandatory parameters than on procedures and actual performance, i.e. regulations which prescribe standard analysis/project procedures encouraging designers to define and implement repair/reinforcement actions in line with techniques and other features typical of the context in which the buildings concerned are located;
- An overall LEGISLATIVE FRAMEWORK which makes for the launching of permanent maintenance schemes geared toward the adoption of anti-seismic reinforcement techniques consistent with the characteristics of the local historic architectural heritage;
- TRAINING SPECIALISTS which will act as facilitators in on-site actions aimed to reinforce historical buildings through an intelligent use (and up grading) of traditional anti-seismic techniques and train local technicians in the relevant activities.

Finally, a last remark. The exclusive focus on historical buildings is only apparently a limitation. As mentioned in the introduction, most of the victims of an earthquake die under buildings collapsing. In earthquake regions most buildings pre-date the widespread introduction of reinforced concrete. Their vulnerability is a direct function of the seismic culture - conceived of as a combination of technical know-how and modes of conduct consistent with it - of the community which originally erected them and has continued to use them ever since. As these repair and reinforcement actions are often assigned to technicians who are not university graduates -

or to graduates not having a specific knowledge on ancient and non engineered built-up - without they may greatly affect the stability of the historical built-up heritage.

In this context, to use the LSC approach can help to make the anti-seismic prevention policies more effective.

REFERENCES

Cardona O. 2001. *Structural analysis of historical buildings* in *Old buildings and earthquakes*. Strasbourg, Council of Europe and European University Centre for Cultural Heritage

Giuffré A., 1992, Sicurezza e conservazionbe dei centri storici in area sismica. Sintesi metodologica. In *Sicurezza e conservazione dei centri storici in area sismica. Il caso di Ortigia*, edited by Bari, Editori laterzaTouliatos P. A..

Hu Shiping. 1991 "The earthquake-resistant properties of Chinese traditional architecture." *Earthquake Spectra* 7/91.

Tanabashi. 1960. "Earthquake Resistance of Traditional japanese Wooden Structures." In the Proceedings of the *2nd World Conference on Earthquake Engineering*, Tokyo.

Toulaitos P. 1993. "Traditional aseismic techniques in Greece." In the Proceedings of the Interantional Workshop *Les sytèmes nationaux faces aux seismes majeurs*, edited by L Mendés Victor, Lisbon, EUR-OPA Major Hazards.

RECENT POLICY DEVELOPMENTS IN TURKEY

9

Structural and Educational Dimensions of Earthquake Damage Mitigation in Turkey

Tuğrul Tankut
Middle East Technical University
Ankara, Turkey

1. INTRODUCTION

Earthquake hazard mitigation involves a variety of activities covering a wide range extending from engineering to administration, from education to finance, from earth sciences to social sciences. Areas concerning mitigation activities can perhaps be classified into four major categories.

1.1 Natural Phenomena

A clear understanding of the seismicity of the entire country, particularly regions of high seismic risk, based on reliable earth sciences research is naturally a very important component of earthquake mitigation.

1.2 Built Environment

Earthquakes cannot be prevented, but their damaging effects can be minimized by making the structures seismically safe. The present earthquake engineering state-of-the-art enables seismically safe design and construction of residential, commercial, industrial etc. building structures as well as infrastructure systems such as transportation, power, water etc. To achieve a seismically safe built environment, the existing structures, which do not comply with the present seismic requirements, should be strengthened, while all the new structures should be designed and constructed in accordance with the current Seismic Code requirements.

P.R. Kleindorfer and M.R. Sertel (eds.), Mitigation and Financing of Seismic Risks, 179–189.

1.3 People

If the people are intellectually and psychologically prepared to face the earthquake on rational grounds, and if they realize the importance of having their buildings evaluated and strengthened for seismic action, earthquake hazard mitigation can successfully be realized. Public awareness education is therefore essential.

1.4 Administration

A well organized and efficient disaster management system, systematically operating before, during and after the earthquake is another very important component of effective earthquake mitigation.

Only two important issues related to two items, built environment and people mentioned above, namely the structural and educational aspects of the problem of earthquake mitigation is discussed briefly in the present paper, in the light of observations made during the recent five urban earthquakes. Actions to be taken in these two areas towards effective earthquake preparedness are presented in a classified manner and some proposals concerning these actions are made.

2. OBSERVATIONS IN RECENT EARTHQUAKES

The three major and indisputable observations made during the recent urban earthquakes in Turkey were (i) excessive structural damage, (ii) inefficient disaster management and (iii) irrational interpretation of earthquake.

2.1 Excessive Structural Damage

The structural damage in all the recent seismic events, considering the magnitude of the earthquake, was much heavier than one would normally expect in a country better prepared for earthquakes. The author is convinced that the causes of this severe damage can be classified into three major categories.

2.1.1 Deficient Engineering Practice

Leave the unengineered buildings aside, even the majority of the engineered residential low-rise and mid-rise buildings do not comply with the seismic code of their time of construction. This deficient engineering practice may appear surprising, if the top quality performance of the Turkish construction industry competing successfully in the international market is considered. This is the Turkish reality; beside common sub-standard applications, one can always find very successful examples, in almost every area.

2.1.2 Ineffective Supervision & Inspection

Another very important cause of heavy damage is the ineffective and distorted supervision and inspection system in use. Inspection of the design and construction of residential buildings in urban areas is performed by municipalities that, in many cases, do not employ any qualified engineers, sometimes any technical personnel at all. Most of the municipalities have totally ignored the Seismic Code for some time in the past, considering it too complicated to be enforced. Some of the decisions of some municipalities concerning land use and development were based on political pressure and personal benefit considerations rather than technical recommendations.

2.1.3 Insufficient Demand for Seismic Safety

The author is convinced that, if the people realize the vital importance of the seismic safety of their homes and if they are prepared to pay a small cost (no more than they are willingly paying for the colored bathroom tiles) to buy seismic safety, the construction industry will somehow meet that demand. In an effort to maximize its profit, the construction industry did not pay much attention, in the past, to seismic safety, which did not appear to sell for an attractive price, and could somehow manage to avoid its legal and professional responsibilities in this regard.

2.2 Inefficient Disaster Management

In the modern sense, the concept of disaster management involves not only the post-quake activities such as rescue, aid, sheltering etc. operations, but also long term activities towards a comprehensive planning and a thorough preparedness scheme. Such a disaster management system could not yet be established and operated in Turkey. Very effective and useful activities of the Crisis Management Center in the Prime Ministry and the local centers directed by the governors of the disaster areas were all confined to the post-quake operations, and therefore they were all far from the modern definition of the concept.

2.3 Irrational Interpretation of Earthquake

With the exception of the well educated section, most of the people take earthquake as a natural disaster (implying that it is an act of God), which should be accepted without questioning. These people do not realize that (i) earthquake is a natural phenomenon (just like gravity) and one can survive with it by taking the necessary measures; (ii) their homes can be made seismically safe and its cost is not very high; (iii) however, the cost of failure which may include their own or their children's lives is incomparably high. Furthermore, they realize neither their rights to demand justice for the engineering deficiencies causing them enormous losses, nor their duties to have their buildings seismically evaluated and retrofitted if necessary, and to have them covered against natural disasters by insurance. Instead, they expect the state (ironically called "father") to take care of everything and to recover all their losses. Indeed, the 62 years old "Disaster Act", passed in the age of a very strong state controlled economy and still in force in the present medium of market economy, requires replacement of all the earthquake struck homes by the state.

In parallel with the above explained irrational understanding of the ordinary people, some of the engineers also display an irrational approach in their practice. They invariably consider gravity loads carefully in their designs realizing that this kind of deterministic loads will definitely be acting on their structures, whereas they do not mind ignoring the seismic effects hoping that their structures will be lucky enough to escape, during their service life, a damaging earthquake the probability of occurrence of which is rather small. In other words, they gamble with the earthquake.

3. PRINCIPLES OF EARTHQUAKE MITIGATION

The observations briefly explained above are the clues to establish the basic principles of an effective earthquake hazard mitigation system.

3.1 Seismically Safe Structures

The first and probably the most important requirement of coexistence in peace with the earthquake is to have seismically safe structures, including both buildings and infrastructure systems. The losses will be minimal, if both the existing structures and the new ones to be built can somehow be given the required seismic safety.

3.2 Efficient Disaster Management

The need for an efficient disaster management system, based on scientific principles to be established and operated as soon as possible, is obvious. However, this very important problem is beyond the scope of the present paper, and it will be treated by the experts of the area in another session of the workshop.

3.3 Well Informed Public

A systematic public awareness education leading the people towards a rational interpretation of the earthquake related problems and towards realizing their rights and responsibilities concerning earthquake preparedness will no doubt serve the earthquake hazard mitigation purpose very effectively.

4. ENGINEERING AND EDUCATIONAL ACTIONS

The important actions to be taken, concerning the structural and educational issues in the development of an effective earthquake hazard mitigation system are listed and briefly explained in the paragraphs below.

4.1 Seismically Safe New Structures

Strict application of the Seismic Code is essential for the seismic safety of the new structures to be built in the future. Although a rather satisfactory seismic code has been in force since 1975, it is not possible to claim that the code could be implemented in Turkey, and that the structures built in the last twenty-five years are seismically safe. Employment of qualified engineers by the municipalities is obviously essential for the adoption and application of the code in the design and construction of low-rise and mid-rise residential buildings, which constitute the most widespread problem area.

One major factor, which led to the above mentioned situation is, no doubt, the ineffectiveness of the present construction supervision and inspection system. This system should be critically evaluated and restructured. The new construction supervision and inspection system should also include some effective measures to prevent illegal interventions such as bribery, political pressure etc.

4.2 Seismic Retrofitting of Existing Structures

A huge building stock exists in the earthquake prone areas of the country. This building stock can neither be replaced nor ignored; it has to be seismically evaluated and retrofitted. This enormous task necessitates the participation of a large number of practicing engineers. However, seismic evaluation and strengthening are critical operations requiring expertise; they should not be performed by inexpert engineers, no matter how experienced they may be in routine design and construction work. The need for a comprehensive engineer training program is therefore evident.

Besides the existing building stock, the existing infrastructure items, such as bridges, dams, towers etc. also need seismic evaluation and rehabilitation. This is another huge task, which necessitates, in most of the cases, even further expertise compared to building structures.

The use of a practical method of seismic vulnerability evaluation is essential in dealing with the huge existing building stock. Some methods proposed for screening, preliminary evaluation and final evaluation stages can be found in the literature. However, all these methods have been developed considering the common building types, common design and construction practice, commonly used local materials, current legal documents etc. of the country where these methods were proposed. On the other hand, quite satisfactory methods of seismic vulnerability evaluation have been used by the faculty members of some Turkish universities in their rehabilitation studies following the recent earthquakes. However, it is still desirable to formulate a unified seismic evaluation procedure on the basis of local conditions by the consensus of the experts.

Similarly, practical and feasible methods of seismic retrofitting also need to be developed considering again the local conditions. Rather satisfactory methods of post-quake repair, developed, verified and applied by some Turkish universities do not appear to be very suitable for the seismic strengthening interventions on the undamaged buildings, which are still in use. New methods of seismic retrofitting to be developed should not require evacuation of the building if possible, and should cause minimum discomfort to the occupant.

It is evident that the realization of this enormous work of seismic vulnerability assessment and retrofitting of the huge existing building stock necessitates the participation of a considerable number of qualified engineers.

4.3 Seismic Vulnerability Assessment/ Certification/ Insurance

The concern caused by the latest earthquakes created a great demand for seismic vulnerability evaluation of the privately owned buildings. Most of the applications are being turned down by the qualified professionals due to their over saturated work programs. Misleading reports are being produced and trivial remedies are being proposed by some unqualified persons in many cases. Seismic evaluation requires a rather high level engineering expertise. A system of "Building Evaluation Centers", employing qualified engineers, licensed after taking a serious course and passing a serious examination, may provide a reasonable solution. The natural solution of this problem would have been authorization of the professional engineers for seismic rehabilitation, if a system of professional engineers existed. Unfortunately, this is not the case in Turkey; the young engineer assumes full authority the day he/she receives his/her diploma irrespective of the level of education he/she received.

The system of building evaluation centers can easily be associated with a system of "Building Certification" which can operate on either voluntary or compulsory basis, to classify the seismically evaluated buildings into various grades (A, B, C etc.) indicating their present seismic safety level and the need for seismic strengthening. Such a certification will no doubt affect the market value of the buildings and thus encourage the property owners to have their buildings evaluated. Furthermore, it can also be related very suitably to the earthquake insurance mechanism, which in any case requires seismic assessment of the buildings to be covered for a rational risk evaluation and premium determination.

Insurance with seismic coverage is no doubt another incentive to promote seismic safety. However, it is beyond the scope of the present paper, and it will be treated by the experts of the area in other sessions of the workshop. Nevertheless, the author is somewhat pessimistic about the success of such an insurance system, as long as the "Disaster Act" remains in force. Because, this act is nothing but a state guarantee for replacement of the earthquake struck homes; in other words, it is a kind of insurance free of charge.

4.4 Land Use and Development

It is hard to claim that any consideration has been given to potential disasters in the town and regional planning work performed until very recently. Municipalities contribute to (if not control) the preparation of the land use and development plans, and they have full authority to execute these plans. Considering the politicized nature of municipalities and their lack of technical personnel, it should not be very hard to imagine on what grounds the land use and development plans are often modified. Some technically very unsuitable areas may sometimes be designated as residential or commercial development areas; sometimes technically very unsuitable type and size of buildings may be permitted etc. on the basis of political pressure and personal benefit considerations.

In the future town and regional planning work, a disaster sensitive approach should be adopted. In the preparation of these plans, not only the earthquake hazard, but also the other potential disasters should be given serious consideration. Disaster sensitive revisions of the existing plans and the land use and development maps are also necessary.

A distorted version of the amnesty concept is a Turkish invention. Occasional amnesties (mostly on political grounds) legalize unlawful land use and technically unacceptable and therefore unsafe constructions and town development; just like the educational amnesties which bring the dismissed students back to the university over and over again. From the earthquake hazard mitigation point of view, it is imperative that such amnesties should be avoided in the future.

4.5 Engineering Education

Poor engineering practice appears to be one of the major factors behind the severe structural damage leading to great losses. It should be confessed that the current civil engineering curricula are deficient as far as the earthquake problems are concerned. This problem of engineering education is twofold; future engineers need to be equipped with a better earthquake background; the earthquake deficiency in the background of the practicing engineers should somehow be compensated.

Departmental curricula in Turkish universities consist of two components; the core curriculum determined by the Higher Education Council, plus the compulsory and elective courses left to the discretion of the university. The civil engineering core curriculum does not contain any earthquake engineering courses; such courses are offered in some universities only on elective basis. The core curriculum needs a careful revision from this deficiency point of view. Besides, each university should reconsider its educational objectives and review its civil engineering curriculum carefully. In the revision, a special attention should be placed on the inclusion of engineering ethics, which appears to the author as another major deficiency of the present curricula.

Although it is possible to specialize in the area of earthquake engineering in structural engineering graduate programs of a few universities, it is advisable to launch an earthquake engineering (and possibly also a rehabilitation engineering) graduate program in some universities where adequate faculty exists. The subject definitely deserves further attention.

The number of state and private universities and consequently the engineering programs increased enormously in the last few decades. A sub-standard engineer training appears to be unavoidable in these developing universities due to lack of competent faculty. Since a system of professional engineers does not exist in Turkey, the engineers graduating from these universities with a questionable background are given full authority. A considerable contribution of these engineers to the poor engineering practice is estimated. The policy of employment of competent faculty in these newly founded universities is a definitely required long-term remedy. However, introduction of system of professional engineer in which a certain experience and exam proven qualification are required for engineering practice with full authority, can no doubt improve the standard of engineering practice in the country.

On the other hand, various types of compensation courses are definitely needed for the practicing engineers to familiarize them with the problems of earthquake engineering. Some of these courses must be certificate courses where attending engineers may be licensed to undertake seismic evaluation, repair and strengthening works, following a serious examination. Besides, some refresher courses should be organized regularly.

4.6 Public Awareness Education

Public awareness is one of the most important components of earthquake preparedness and consequently a very effective factor in coping with the disaster. It is a lengthy operation, which takes insistence, consistence and plenty of patience.

All possible ways and means should be used for this purpose. However, the media is probably the most effective means to convey some very important messages to the public such as:

a. Earthquake is a natural phenomenon and one can survive with it by taking the necessary measures;

b. Your homes can be made seismically safe and its cost is not very high;

c. However, the cost of failure which may include your own or your children's lives is incomparably high;

d. You have right to demand justice for the engineering deficiencies causing you enormous losses;

e. However, you also have your duties to have your buildings seismically evaluated and retrofitted if necessary;

f. And to have them covered against natural disasters by insurance.

Giving advice alone is not enough; these messages need to be supported by convincing material and especially by reliable organizations. For example, when someone preaches the importance of having the buildings seismically evaluated by experts, he should be able to refer the audience to a building evaluation center where the work can be done properly, without much delay and at a reasonable cost.

5. EPILOGUE

This overview paper presents a brief summary of the multidimensional actions of a very wide scope to be taken towards developing an effective earthquake hazard mitigation system in Turkey. None of the points raised in the present paper is new; all these points have been proposed and discussed for a very long time in scientific and technical meetings by various experts of the area. However, these simple and straightforward proposals could never be realized, mainly due to the lack of interest in the authorities who in general prefer to spend the public funds in post-quake activities (soothing the pain), which can easily be noticed by the electorate.

If some of the proposed actions were started earlier, and if just a fraction of the funds spent after the earthquakes was allocated for the preparedness activities prior to earthquakes, the enormous structural, economical and moral damage caused by the recent earthquakes could have been significantly reduced. However, everybody seems to have learnt a lesson from the latest disasters. Although he is still afraid of the present awareness to fade out in time, the author wishes to be able to hope that some of the necessary measures will be taken or at least some serious work will be started this time.

Better late than never!

REFERENCES

Tankut, Tuğrul. 1999. "Earthquake Problem in Turkey – TUBITAK Perspective", report presented at the Earthquake Working Group meeting, NATO Science and Environmental Affairs Division, (December), Brussels.

Tankut, T., Duman, Ş., Yılmaz, R., Duyguluer, F., Demirtaş, G., Soğancılar, E. 2000. "Natural Disasters Working Group Report" (in Turkish), Eighth Five-year Plan preparatory work, State Planning Organization, (April), Ankara.

Tankut, T., Ersoy, U., Ergun, U., Koçyiğit, A., Gündoğdu, N. 1992. "1992 Erzincan Earthquake – Reconnaissance Report" (in Turkish), TUBITAK Construction Technology Research Grant Committee, (November), Ankara.

Tankut, Tuğrul. 1999. "1999 Marmara Earthquake – Reconnaissance Report" (in Turkish), TUBITAK Construction Technology Research Grant Committee, (August), Ankara.

Tankut, Tuğrul. 1996. "Importance of Public Education in Earthquake Preparedness" (in Turkish), TUBITAK Earthquake Engineering Symposium, (February), Ankara.

Tankut, Tuğrul. 1997. "Action Plan for Seismic Safety Evaluation of State Owned Buildings in Istanbul", Fourth National Conference on Earthquake Engineering, (September), Ankara.

Özateş, Balkan. 1999. "Seismic Evaluation – A Method Proposed for Final Evaluation of Reinforced Concrete Buildings", MSc thesis, Middle East Technical University, (November), Ankara.

10

Revision of the Turkish Development Law No. 3194 Governing Urban Development and Land Use Planning*

Polat Gülkan
Earthquake Engineering Research Center and
Disaster Management Research Center
Middle East Technical University
Ankara, Turkey

INTRODUCTION

This paper is based on a report[1] that stemmed from an investigation into improving Turkey's legal framework for spatial planning and physical development. The principal motivation for the investigation has been the renewed awareness in the wake of the Erzincan earthquake of 13 March 1992 that there exist systemic defects in the way the built environment in Turkey is created. These deficiencies cause the building stock to have poor record against disasters, and bleed the national economy. It drains resources in an endless cycle of rebuilding after each occurrence of a disaster. In view of the great losses from the 17

---

* The lengthy title of our report is: "Revision of the Turkish Development Law No. 3194 and Its Attendant Regulations with the Objective of Establishing a New Building Construction Supervision System Inclusive of Incorporating Technical Disaster Resistance-Enhancing Measures." The consulting services under the scope of this investigation were supported by the Turkish Government Housing Agency through the General Directorate of Technical Research and Implementation, Ministry of Public Works and Settlement. The research team included Murat Balamir, Haluk Sucuoğlu, Raci Bademli, Bengü Duygu, Melih Ersoy, and Gönül Tankut. This paper reflects our collective views.

1 See Note *

P.R. Kleindorfer and M.R. Sertel (eds.), Mitigation and Financing of Seismic Risks, 191–206.

August 1999 and 12 November 1999 earthquakes in northwestern Turkey, parliamentary adoption of the types of legal and structural instruments that have been developed during the course of the investigation has become more pressing.[2]

The Final Report was based on the principal issues identified during the two earlier phases containing the draft texts for the proposed legal changes. The two reports formed the background feedback received during the national consultative meeting held in late February, 1999. The Final Report, submitted in draft form on 10 August 1999, describes revised or new institutional and technical instruments for their solution. Draft texts for revised laws and regulations addressing the deficiencies identified have been appended to that report. A number of new laws and regulations have been formulated. Many of the suggested revisions have already been enacted into law, albeit with shortcomings (Gülkan, 2001).

1. BACKGROUND

Turkey's long history of disasters, most frequently earthquakes, led in 1958 to the establishment of a Ministry of Reconstruction and Resettlement. The ministry was responsible for the implementation of the Development Law and the Disasters Law. The primary objective in setting up the ministry and its agencies was to reduce the risk of death and injury to the population, and as a second but equally important priority to reduce the scale of the economic risks involved. The Ministry was made responsible for updating and promulgating both the seismic building code and the earthquake-zoning map. (The latest revision of the code became effective as of 1998, and the map in 1996.)

Until 1985, when the current Development Law was enacted, urban planning departments in Turkey were part of government offices representing the Ministry of Reconstruction and Resettlement. The role of the authorities included land use designation (the preparation of a land use master plan), control and compliance with zoning ordinances, licensing new developments by private owners, and locating public facilities. From 1985, these privileges were transferred to the local governments. The Development Law has the declared intention of controlling the appropriate formation of settlements and buildings. The mission of controlling only

[2] The Government of Turkey adopted a decree on 27 December 1999 for the establishment of a compulsory natural disaster insurance pool. A new entity called "Emergency Management Authority of Turkey" has been created. In April 2000, a law for building construction supervision and professional liability was passed. These are all positive steps, although they are not as comprehensive as our proposals.

the construction phase is narrow, because it excludes organization of investments and entrepreneurship, provision of land and other infrastructure, technical means of oversight during the construction. Property management approaches and the protection of the various kinds of the environment are also not included in the scope of the law. Powers of plan making and ratification have been delegated to the local governments, irrespective of size and manpower resources. To compound the difficulties, other ministerial bodies have also been entrusted with plan making powers (Gülkan, et al., 2000).

Control over enforcement of building codes in privately owned buildings is possible within municipal bounds, defined loosely as townships with more than 2000 population (there are at present some 3200 municipalities in Turkey) where municipal engineers theoretically have powers to enforce compliance with regulations. Building plans are submitted to the municipal authorities with the signature of a design engineer who is responsible for code compliance. In practice, municipal engineers are not able to check thoroughly all of the design calculations because of their heavy workload. The number of provinces and municipalities arranged on the basis of the highest hazard zone within their boundaries is shown in Figure 1.

Figure 1. Distribution of Provinces and Municipalities among Seismic Hazard Zones

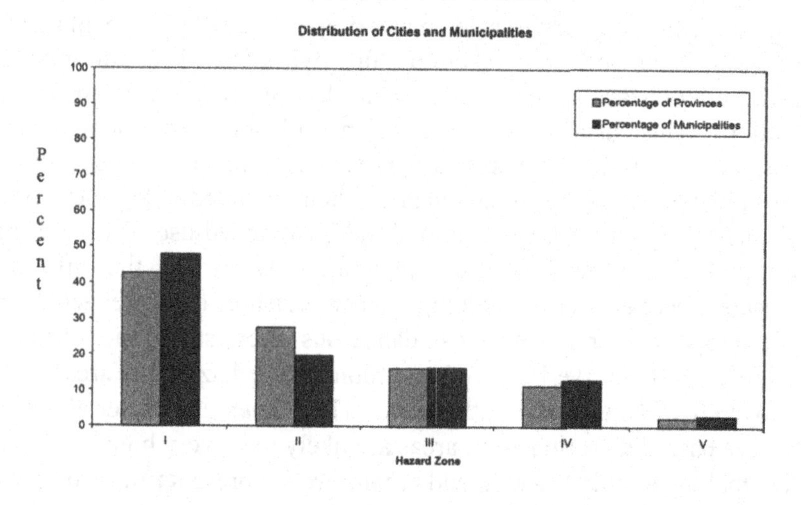

The layout and development of cities, location of the infrastructure, key buildings and utilities, and the physical development of the built environment can all affect the impact of an earthquake. The town planner, the regional planner, the engineer designing the layout of the utility networks, transportation routes or key installations, and anyone whose job is to locate facilities within a town or whose decisions affect the use of the land, all have a role to play in reducing the potential impact of, say, an earthquake (Coburn, 1995).

Effective management of urban development in a city depends on understanding the processes that shape it. The trends in land prices, location preferences for different types of industrial facilities, community development patterns, and population growth are the complicated dynamics that shape a given city. Urban planning attempts to regulate these processes using the powers of legislation and economy. The principal concerns of town planning are directed toward the creation of a safe, clean and pleasant environment. Disaster management concerns follow a parallel course: limiting the densities of population and development, provision of critical services at times of emergencies, and ensuring the continued operation of economic activities. Urban planning is a long-term process, therefore disaster management and assets protection are also long-term (Mader, 1997). Traditionally, the disaster-worthiness of the building stock has not been a concern for the urban planner, but of the engineer.

1.1 Enforcement Prospects

In many Turkish municipalities, particularly in those where rapid economic growth has been registered within the last twenty years or so, the zoning ordinances and master plans prepared by the town planning departments have been overtaken by the dynamics of urban growth. Changing circumstances occur faster than planning responses can be put into action. This in effect has resulted in a planning environment that follows, rather than directs, patterns of urban development. Zones defined in master plans cannot be maintained as their intended categories, with many zones being transformed into ill-defined mixed-use areas. Even further removed from the formal planning process are the informal settlements where almost no building quality measures can be enacted. In many metropolitan areas the most dangerous sites, steep and unstable hills, stream gullies, riverbeds and environmentally hazardous areas have been covered with runaway settlements. The human and material losses of a severe hazard affecting these areas are likely to be very high.

Plan making at only regional and urban levels represents an incomplete hierarchy. The Ministry intervenes as an intermediate authority if conflicts arise between local authorities or in cases when national concerns are involved. Land preparation, sub-divisioning, and rearranging property rights in areas where rural to urban conversions occur are the basic operational tools of the law. Building permits are obligatory, designs being submitted to the local governments to meet the requirements of both building and disaster regulations. The actual construction activity cannot be supervised too well because of inadequate personnel and financial resources. Yet, local governments are responsible for this supervision and the issuance of occupancy permits.

The scope of the Disasters Law dating from 1959 is to provide a public intervention capacity and improvement in the efficiency of relief operations after disasters. These operations are entrusted to provincial governors who are granted wide powers in the aftermath of disasters. Plans of disaster-affected settlements are immediately revised and construction permits are granted. Settlements that have been destroyed substantially may be relocated by decree of the Council of Ministers.

Apart from its restriction to post-disaster operations, and its independence from the Development Law, the Disasters Law falls short of being a contemporary disaster management blueprint. One of its most controversial articles is that every homeowner will have a dwelling built with public funds if his former property is unusable after the disaster. This stipulation does not differentiate between authorized and unauthorized construction. Building owners who abide with the legal requirements, and those who insure their property are effectively penalized by the system because they are excluded from the handsome subsidies meted indiscriminately to other "entitlement holders" from taxpayers' funds. It is acknowledged that this not only becomes a source of injustice, but contributes also to a culture of fatalism, leading to the creation of a society in expectation of disasters. With the above review the components of a strategy to transform the system to a model where a prepared, rather than a fatalistic, society can be recognized as a priority (Balamir, 1999).

Major legal restructuring is necessary. First, the Disasters Law should be stripped of its pretensions of disaster preparedness, and these functions should be embedded in the Development Law itself. The former law needs to have its post-disaster procedures revised, with particular reference to re-defining the terms and conditions of entitlement rights. No individual should be entitled to public subsidies without proof of full compliance with the ordinary development constraints. Such entitlement could be based on a registered qualification of property, leading to differentiated property markets at the early stages of property valuation. Further improvement will be the encouragement of individuals for buying out building insurance, and the transfer of a part of property taxes collected by the municipalities to the Disasters Fund. Differential property taxes ~~could~~ might be levied by local governments to penalize unauthorized development.

Amendments in the Development Law must include the following headings: a general upgrading of planning supervision, and unification of powers for planning with a comprehensive hierarchy of interrelated plans. Incorporation of participation, community protection, urban renewal and design and property management will not only complement the existing planning functions, but also improve the background for disaster management operations.

The principal instrument governing how buildings are created is the Development Law. This document has a few articles in Part 4 that regulate the supervision of building construction. The law holds municipalities (or governorates for buildings outside of urban areas) responsible for project supervision. Construction supervision is entrusted to the so-called technically responsible engineers.

Holders of deeds or parcel assignment certificates submit petitions to either the relevant municipality or the governorate to acquire building permits. In addition to the certificate of land ownership the applicant must submit architectural, structural, and mechanical designs as well as a schematic drawing of the buildings location. Some municipalities have transferred this duty to the local branches of the Chambers of Civil Engineers or Architects through informal agreements, but this practice is questionable because the law clearly holds the local government liable for ensuring the life and property safety of the people it serves. The customary procedure is that the technical offices of municipalities function as rubber stamps in their approval work. The Development Law does not specify what measures are to apply if erroneous designs are approved. Sparse legal precedent appears to hold the design engineer responsible in this regard.

The Development Law requires the technically responsible engineer to report to the municipality or governorate any contraventions by the contractor of the design he supervises. When such a violation occurs it is incumbent upon the local government to seal the construction site, and to order the owner to take corrective action. If within one month this action is taken, the order for work stoppage is rescinded. If the owner does not comply with the order, then his permit is revoked, and the building demolished at his expense. This process is largely illusory.

Limitations exist in how effectively the land-use planning instrument can be used as a tool for disaster mitigation. Land-use planning is an opportunistic activity because it can be exercised only in the case of an expanding city where there are choices between alternative areas. Land-use planning must be controllable, and this is where the greatest difficulty exists within the Turkish system. Without major changes in the planning control mechanisms currently in use according to the Development Law, detailed control over private urban development will remain illusory. A less tangible factor in shaping cities is the price of land. Hazards can influence the price of land, and change the shape of a city. For example, riparian land where ground motions are more severe than on firm ground can potentially suppress land prices there in spite of scenic advantages it may offer. Theoretically, an educated real estate market in a rich economy can make proper decisions to protect itself against hazards (Johnson, 1998). This is not yet within reach in Turkey, so strict enforcement is necessary. For these measures to have any success there must be a social agreement for the communal benefits they will engender, and a willingness to pay the corresponding price.

It is not clear what quantifiable reductions in disaster potential will accrue if changes in land use are enacted. This could be attempted on a theoretical basis by determining the optimum location of facilities where location affects seismic vulnerability. The change in damage potential is connected to changes in subsoil or distance from active faults. It is then possible to calculate the added benefit of either relocating facilities or of strengthening them taking into account the special requirements these conditions may impose. The effectiveness of different schemes for above-code measures for strength could then be compared with the returns they provide. The most sensitive item in such assessments would be the quantification of the lives, and the economic value saved.

Any policy recommendations following from mitigation needs must carefully balance costs and the accruing benefits. It would be very useful if a quantitative framework could be drawn for defining risks, vulnerabilities, and mitigation actions and their costs. To my knowledge such work has only been done to date only on a theoretical basis, with no physical calibration because the answers to these questions depend on many social and economic considerations where formal decision making techniques cannot be applied directly. Even with uncertainties, the quantification of costs and the corresponding benefits for earthquake protection measures can illuminate the decision making process. Loss estimates should be of much interest to those who do physical or economic planning because planning decisions can have an effect on future losses. We will refrain from entering the subject of loss estimation. In the Appendix, an answer is given to the hypothetical question of what savings might have been possible in Dinar if a pre-emptive building rehabilitation program had been undertaken there prior to the 1 October 1995 earthquake.

It is acknowledged that a revision of Turkey's disaster management capacity is necessary. Within a cost-sharing arrangement with the UN Development Programme, the Ministry of Public Works and Settlement is currently executing a program entitled "Improvement of Turkey's Disaster Management System." This program aims at institutional strengthening with carefully designed seed projects, each addressing a specific area for which a given agency is responsible. Additionally, a more ambitious and comprehensive program (code-named TEFER: Turkey Emergency Flood and Earthquake Rehabilitation) is underway to meet this objective. TEFER[3] is funded by a loan from the World Bank.

Reducing the vulnerability of existing buildings is an important aspect of any earthquake hazard mitigation program. Maintaining the existing stock is necessary not only for economic but also social and cultural reasons. In many areas the older, weaker buildings where poorer people live are the sources of expected future losses. Nearly all options for strengthening are expensive when compared with the incorporation of capacity building elements into a new design, so the costs and benefits must be carefully analyzed in deciding whether to strengthen or to demolish. In areas of higher seismic risk, upgrading is more likely to be more cost-effective because the cost per life saved, or per saved time of economic disruption is lower (Coburn, 1995). Unfortunately, such rational analyses have not been performed prior to decisions leading to the rehabilitation schemes in Turkey because appeasement of public suffering has always had top priority.

1.2 Deficiencies of the Current Development System in Its Powers of Mitigating Disaster Effects

The planning system with its numerous regulatory mechanisms and actors is far from a unified and singular body or authority of monitoring physical development in Turkey. Apart from the local authorities, separate and distinct powers of development planning, plan ratification, building construction, are currently enjoyed and extensively exercised by a multitude of ministries and other public bodies. This brings vagueness in the identification of responsibilities and weaknesses in the control of development in an area demanding strict discipline in conduct and clarity and coherence in authority. The responsibilities of enforcing the Development Law No. 3194 lies with the Ministry of Public Works and Settlement. The Ministry itself is deprived of all means of control over municipalities, the real beneficiaries enjoying the powers given by the Law in their daily practices. All powers concerning inspection of the municipalities on the other hand, are delegated to the Ministry of the Interior which in turn has no technical capacity to control plan-making and building.

The Development Law is solely a physical regulation instrument for development. It ignores the finance, organization, protection, and

[3] Following the 1999 disasters, TEFER has been linked to another World Bank program code named MEER: Marmara Earthquake Emergency and Reconstruction.

management issues that are integral or reciprocal requirements of all building activity. It has little power or incisive tool to manipulate or physically rearrange properties (and the rights of ownership), to maintain the 'public welfare', particularly monitor building activity in disaster areas. The Law has no provision to cope with natural (and other forms of) disasters in itself, neither does it have an authentic interrelation with the Disasters Law No. 7269, apart from minor referencing between their respective by-laws.

The Disasters Law pays only lip-service to pre-disaster preparations and control. Almost all of its provisions are concerned with the aftermath. It disperses public funds without necessarily ensuring their return, which makes it a highly popular mechanism with the political authorities. Throughout the Turkish practice with disasters, earthquakes in particular, powers provided in this Law for technical decision-making has been transferred to the body politic with ad hoc laws passed immediately after each disaster. This form of deprivation of powers of the technical body has been experienced once again after 17 August 1999. The Disasters Law punishes the building insurance policy holders against disasters in its compensation of losses. Furthermore, public credits are benevolently provided to all property owners without discriminating against unauthorized buildings.

All planning and building supervisory action has been charged to the responsibilities of the municipalities (within their jurisdictional borders) and the provincial governors (in all areas external to the municipal boundaries). Even if they intended to, municipalities are in no position to carry out these functions of control simply for the reason of their incapacities in technical man-power, and the sheer size of the tasks involved. As if to complement this, municipalities are not liable, to any serious extent, for their remission of responsibilities of development control. This situation is aggravated by the fact that no legal countermeasures have been spelled out for those officials who fail to perform their duties in the spirit of the law.

1.3 Deficiencies of the Current Building Construction Supervision System in Its Powers of Mitigating Disaster Effects

The existing Development Law (No. 3194) entrusts the public duty of building construction supervision to the municipal governments. In theory, the design checks for new construction are performed by the building divisions within the city development offices. These offices issue the construction permits, and the daily supervision of the actual construction is performed by engineers or architects appointed by the municipal offices to serve as the watchdogs on behalf of the municipalities. This system of supervision is largely illusory firstly because municipalities do not have qualified personnel in adequate numbers on their rosters with whom to conduct a meaningful design check. With particular reference to implementing the earthquake resistant building code, there exists a great need to employ qualified structural engineers. This is never the case. Municipalities are not legally liable for failing to perform the structural design checks. Oversight in the implemention of the law is within the sphere of the Ministry of Public Works and Settlement, but local governments do not answer to that ministry but to the Ministry of the Interior. Whatever influence the Ministry of the Interior can bring to bear upon the municipalities is sentenced to fall short because of their non-technical character. Our interviews with municipal officials have underscored the fact that municipalities are not really interested in abiding with the limitations of the Development Law, which many consider as an unnecessary straitjacket with no immediate political returns. Municipal assemblies where all decisions in the domain of development and building are taken have become arenas of mutual compromise and averting of eyes among the political parties. Elected local governments find it very difficult to impose any semblance of discipline in this area, leading to a sort of partnership in disregard of duties between owners of property and the municipalities. It is usually the innocent homeowner who ends up paying for this collusion.

The person designated to conduct the supervision of the actual construction is called the "responsible engineer" (or "TUS" according to the abbreviation in Turkish). They usually receive their salaries from the contractor whom they are supposed to be supervising. Many hardly visit their sites. The law only holds them responsible for reporting to the municipal divisons any deviations from the shop drawings. This process is also largely mythical. To compound the situation even further, no legal definition exists for a contractor or the self-builder. The obligations of these persons to the people to whom they sell parts of the property they have developed is defined only within a few obscure articles of the Law of Obligations, i.e., are defined as being purely commercial in nature. When whole buildings containing the developed dwellings collapse during, say, an earthquake and cause loss of life and property, the obligations of developers are limited only by the clause they they have caused these losses "without premeditated malice." This carries no deterring penalties.

2. THE PROPOSED SYSTEM

An effective disaster mitigation strategy must depend on two basic premises: One is the crafting of an effective spatial planning system in which disaster occurrence is considered explicitly as a prime parameter. This includes strict building construction supervision as the other premise. Supervision of plans, the transparency of their preparation, and accountability of public officials reinforce the safety considerations.

The basic objective of the proposed system is to ensure coordination between general settlement policies and disaster policies with emphasis on pre-disaster mitigation and preparedness. It is proposed that integrated disaster maps (data information systems) should form the basis for physical development plans and Disaster Action Programs for hazardous areas and settlements. The title of the Law has been changed to reflect the profound revisions that have been made in its spirit. It now reads as the "Law for Spatial Planning and Development." Its basic features are described in detail below. Other revised laws are also treated under this heading. Revised regulations follow. The global structure of the newly crafted system is illustrated in Figure 2.

Figure 2. Elements of the Proposed System

New Regulations

- Development Supervision Firms
- Disasters Action Fund
- Preparation of Disaster Maps and Their Technical Contents
- Transfer of Development Rights
- Higher Council for Disasters and Development
- Categorization and Registry of Contractors in Private Construction
- Building Construction Supervision Firms
- Uniform Development for Disaster-Prone and Risky Areas

Revised Laws

- Disasters No. 7269
- Turkish Union of Chambers of Engineers and Architects No. 6235
- Pratice of Engineering and Architecture No. 3458
- Spatial Planning and Development

Revised Regulations

- Preparation, Enforcement and Revision of Development Plans and Disaster Prog.
- Development of Areas for Which No Plans Exist
- Urban Renewal Areas
- Shelters
- Uniform Development for Settlements

3. CONCLUSIONS

The continuing process of rural to urban conversion occurring in many areas in Turkey poses a major risk. Reducing the vulnerability of urban settlements would greatly diminish potential losses. A wide range of policies and techniques exists toward this end. The government and local planners cannot unilaterally make cities safe for everyone. It requires communal awareness, motivation and self-protective action by a range of groups including local governments, construction industry, private businesses, insurance companies, professional societies and NGOs. Everyone needs to be educated in converting the built environment to one with enhanced disaster resistance. Planning activities in Turkey should focus on proven techniques for hazard reduction. These include hazard maps, de-concentration and decentralization of key facilities, protective land-use maps and street safety measures. The control of building quality is essential for urban disaster mitigation, and should be addressed as an overall urban protection strategy. This should best be entrusted to private design and construction supervision companies working in harmony with insurance interests. The hindsight figures derived from the rehabilitation program in Dinar demonstrate the cost effectiveness of planning and engineering preventive measures.

REFERENCES

Balamir, M. 1999. "Reproducing the Fatalistic Society: An Evaluation of the Disasters and the Development Laws and Regulations in Turkey." In *Urban Settlements and Natural Disasters*, edited by E.M. Komut, Proceedings of UIA Region II Workshop, Chamber of Architects of Turkey, pp.96-107.

Coburn, A. 1995. "Disaster Prevention and Mitigation in Metropolitan Areas: Reducing Urban Vulnerability in Turkey." In *Informal Settlements, Environmental Degradation and Disaster Vulnerability: The Turkey Case Study*, edited by R. Parker, et al., The World Bank, Washington, D.C.

Gülkan, P., M. Balamir, H. Sucuoğlu, M. Ersoy, B. Duygu, R. Bademli, and G. Tankut. 1999. "Revision of the Turkish Development Law No. 3194 and Its Attendant Regulations with the Objective of Establishing a New Building Contruction Supervision System Inclusive of Incorporating Technical Disaster Resistance-Enhancing Measures." Earthquake Engineering Research Center, Middle East Technical University, Ankara, (in Turkish).

Gülkan, P. 2001. "Rebuilding the Sea of Marmara Region: Recent Structural Revisions in Turkey to Mitigate Disasters," Wharton-World Bank Conference on Challenges in Managing Catastrophic Risks: Lessons for the US and Emerging Economies, January, Washington, DC.

Johnson, L. A. 1998. "Empowering Local Governments in Disaster Recovery Management." In *Lessons Learned Over Time*, Earthquake Engineering Research Institute Publication No. 99-01, Oakland, CA.

Mader, G. G. 1997. "Enduring Land-Use Planning Lessons from the 1971 San Fernando Earthquake," *Earthquake Spectra*, Vol. 13, No. 1, February, pp. 45-53.

APPENDIX: COST OF SUBSTANDARD HOUSING: DINAR CASE STUDY

Over the last twenty-five years, faculty members at METU have conducted extensive field investigations in damage assessment and disaster evaluation following major earthquakes. Until 1992, these field studies were limited to making immediate reconnaissance reports, followed by sporadic involvement in rehabilitation projects, and long-term scientific investigations. The 13 March 1992, M = 6.8 earthquake in Erzincan (640 fatalities) marked a threshold because in its aftermath first institutional and then privately owned buildings were rehabilitated on a large scale. Three national universities, including METU, have served as the principal consulting agencies for the repair and strengthening programs. This practice seems now to have become a matter of government policy because following the 1 October 1995, M = 5.9 earthquake in Dinar (92 deaths), and the 27 June 1998, M = 5.9 event in Ceyhan-Adana (145 losses of life), similar large-scale and costly rehabilitation programs have been undertaken. For a sense of scale, Table 1 lists the documented damage to only the privately owned housing stock during the two 1999 earthquakes.

Table 1. Damage to residential and commercial units

Province	Collapse/Heavy Damage		Medium Damage		Slight Damage		Total
	Home	Business	Home	Business	Home	Business	
Bolu	12,939	2,450	10,968	2,180	11,817	1,564	41,918
Bursa	121	3	569	24	940	68	1,725
Sakarya	23,967	5,069	17,757	3,576	24,423	2,349	77,141
Yalova	13,929	744	14,497	1,159	12,685	1,881	44,895
Kocaeli	32,445	5,367	38,987	5,482	42,154	5,791	130,206
İstanbul	3104	448	14,750	2,428	14,176	1,944	36,850
Eskişehir	80	19	96	8	314	22	539
Total	86,585	14,100	97,624	14,837	106,509	13,619	333,264

Upgrading the lateral load-resisting capacity of an existing, but damaged, building is an arduous task. Precise quantification of the degree of loss of capacity is not possible because of the high degree of indeterminacy. Often, material and workmanship quality is variable (and invariably poor), as-built drawings not available, and foundations inaccessible. In such cases the best policy seems to be to devise simple rules for identification of vulnerable buildings on the basis of readily identified indexes, and to conduct in-depth assessments for those that have been prioritized. By law, only moderately or slightly damaged buildings may be repaired in Turkey. This process is understood to imply the incorporation of structural walls in the case of reinforced concrete buildings or the encasement of masonry buildings within a shotcreted outer shell so that the lateral force capacity meets the requirements of the code. When the damage degree of heavy (including partial or even total collapse) has been accorded to a given building, it is demolished. Fatalities occur almost exclusively in these buildings.

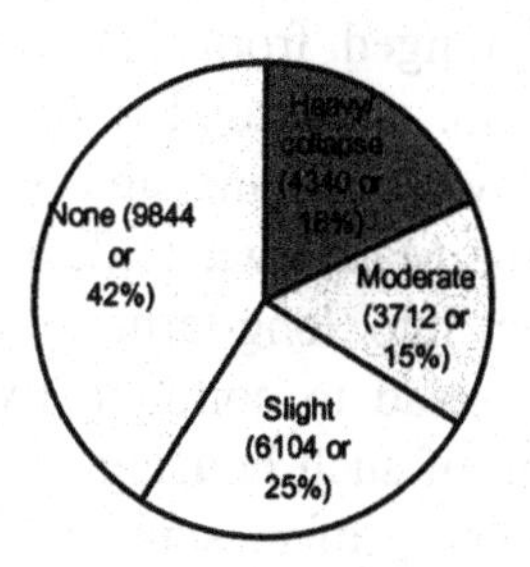

Figure 3. Damage Distribution

Table 2. Classification of Collapsed Buildings in Dinar

Structural System	Number of Stories	Collapsed Buildings Single-Story	Full
R/C Frame	>=4	33	28
R/C Frame	3	29	18
Brick Masonry	4	32	41
Brick Masonry	3	10	4
Composite Masonry	2	6	-
TOTAL		201	

An understanding of the cost effectiveness of repairs in terms of saving lives and preventing injuries can be derived from the data for Dinar. The earthquake on 1 October 1995 caused 92 deaths and about 200 injuries, mostly within the town with a population of 35,000 itself. Sporadic damage and injuries were reported from the villages in the vicinity. The total number of affected people was 100,000, and the number of households 24,000. A building may consist of anywhere from one to several households. For reasons of entitlement canvassing, the Turkish system records only the household information, so building damage must be derived separately. The breakdown of the damage distribution is illustrated in Figure 3. The number of buildings that suffered collapse was 201. Their classification is summarized in Table 2.

For purposes of this discussion we can confine our attention to the distribution of building types within Dinar. The recorded peak ground acceleration of 0.29 g in the city is compatible with a return period of about 100 years, so the levels of moderate damage, implying structural distress and collapse (causing death and injury) correspond to unacceptable response. Ideally, had a pre-earthquake assessment been made, these building would have been indicated as requiring structural intervention for upgrading purposes. Multistory reinforced concrete buildings were hit harder, although they constituted only 15 percent of the building stock. These buildings contributed more household units, and larger total area than their share of the building types. In a simplistic approximation, we may assume that one-third of all buildings, and two-thirds of households needed prior upgrading so that no injuries and deaths would occur. (As stated earlier, upgrading is costlier than full enforcement of code requirements and building quality assurance during construction. This has been calculated to amount to about 10 percent extra cost.)

The METU inventory contained 35 reinforced concrete buildings with a combined floor area of 24,200 sq. m and about 300 household units. The average floor area for a building was 700 sq. m. The repair expenditure varied between buildings, and ranged from $15/sq. m to $63/sq. m, with an average figure of $45/sq. m. The total floor area of reinforced concrete buildings examined also by other teams and requiring structural intervention was about 70,000 sq. m. Many owners decided not to repair their homes, and refused to enter into long-term obligation arrangements with the Ministry, choosing instead to continue living in sub-capacity buildings. About 100,000 sq. m of such housing existed in Dinar prior to the earthquake. Under the category of masonry buildings, including hybrid construction, the METU team decided to repair 79 of the 152 structures for which detailed assessments were made, and the remainder was condemned to demolishment. The total floor area of the 152 buildings amounted to 20,500 sq. m, for an average figure of 135 sq. m for each. The total number of buildings in the masonry classification was 430, with a combined floor area of 60,000 sq. m. Again, there was great asymmetry among the estimated repair expenditures, with an average figure of $30/sq. m. As a guiding figure, it may be assumed that of those eligible only about half were included in the rehabilitation program.

These figures state that, if a pre-emptive structural intervention program had been initiated in Dinar prior to the occurrence of the earthquake, the cost would have amounted to $8,100,000. This figure corresponds to $90,000 per life saved, or about $30,000 per injury avoided. These figures require qualification. In Dinar, precursors had occurred so that most people were already alerted to the possibility of an earthquake. The time of the main shock was about 6 pm, so that most people were still outdoors. In comparison with Ceyhan where on 27 June 1998, 12 buildings collapsed killing 89 people, the number of deaths per collapsed building was lower in Dinar.

The current average cost for reinforced concrete construction in Turkey is $250/sq. m, and for masonry $200/sq. m. If strict quality assurance is assumed to increase these amounts by 10 percent, then a total of $4,900,000 would have been needed prevent the deaths in Dinar, for an average of $53,000 per fatality. These figures may be compared with the total of $250 million, estimated as the total loss bill for the Dinar earthquake.

11

Recent Changes in Turkish Disasters Policy:
A Strategical Reorientation?

Murat Balamir[†]
Middle East Technical University
Ankara, Turkey

1. BACKGROUND

Early morning at 03:15 on the 17th of August 1999, the devastating 45 second 7.4 Richter value Marmara earthquake took more than 18,000 lives in Western Turkey, the more developed industrial heartland of the country.[1] The event left 300,000 dwelling units and more than 50,000 business premises in debris, forcing a population of nearly 600,000 to seek emergency shelter. A second traumatic incidence of similar magnitude took place a step further east on the same fault line, only three months later on the 12th of November. The estimated losses are around 7-8 billion US$, more than a third of the annual total GNP of Turkey. Human suffering, social and psychological impacts of these events have been deep and lasting. The respectability of the public authorities was impaired, and the interests of the industry were seriously damaged, unlike many of the previous occurrences of equally grave disasters that took place at the distant eastern provinces. This generated a strong national consensus and will-power to devise new and effective methods of tackling with disasters. Since then, much effort and debate has been taking place in the political, official and

---

[†] M. Arch., METU; Dip. Trop. Architectural Association London; M. Phil. in Planning, University College London; Ph. D. in Political Science and Public Administration, Ankara University; Associate Prof., City and Regional Planning Dept., Middle East Technical University, Ankara; Member of Executive Board of METU Disaster Management Centre; Member of the National Earthquake Council.

[1] An inventory of the damages experienced in the NE Marmara Earthquake is available at the appendices of the World Bank report no. 19844-TU (1999).

P.R. Kleindorfer and M.R. Sertel (eds.), Mitigation and Financing of Seismic Risks, 207–234.

academic circles to refresh the attitudes, methods of management, the structure of responsibilities, and revise the related legal framework.

A major research financed by the World Bank in the context of the 1992 Erzincan earthquake recovery plan, and commenced three years ago, aimed to revise the very system of urban planning and building control in Turkey. It meant to move away from crisis management *per se*, and devise new methods of safety in planning and pave way to a more prepared society in tackling with the effects of natural disasters.[2] The findings and recommendations report of the final stage of this research was submitted to the Ministry of Public Works and Settlement (MPWS) on the 10th of August 1999, a week before the earthquake.

The calamities of 1999 determined also the fate of this research report, and instead of slipping into oblivion in the shelves of the MPWS, as usually the case is with many of such reports, most of its propositions were addressed in the government's decisions, even if dismantled from their entirety and largely distorted in content and purpose. This report seems to have been the source of inspiration up today, of several pieces of regulations, and many Decrees of the Board of Ministers based on a law (4452 dated 27.08.1999) that empowered them. These covered the 'obligatory building insurance' law, the law to institute 'building control firms.' A more recent Decree regulated the rights and liabilities of practice in the building professions, to complement the series. Still other regulations are promised to follow, and a draft proposal for a new Development Law is circulating. These decisions may be interpreted as attempts to convert the existing system over-occupied with crisis management, into some form of a commitment to an overall strategy for disaster mitigation. It is worthwhile therefore to investigate the nature and likely level of success of these moves in mitigating disaster losses, as well as the deviations from

[2] My intensive involvement with the issue of restructuring existing legal provisions for the mitigation of natural disaster losses owes to the approval of our proposal for a research in 1997. The World Bank supported research was tendered and monitored by the Housing Development Administration under the auspices of the Ministry of Public Works and Settlement (MPWS) and focused on 'The Revision of the Urban Planning and Building Control System in Turkey for the Mitigation of Disaster Effects.' The research was carried out in the context of METU Earthquake Research Centre by a group of engineers, urban planners and architects, mostly academic members of the Middle East Technical University (P. Gülkan, H. Sucuoğlu, M. Ersoy, R. Bademli, R. Keleş, G. Tankut, A. R. Aydın). I was responsible for writing and editing of the chapters on the planning system. The final report of this research renewed the Development Law together with a dozen of bylaws, and was submitted to the MPWS on 10.08.1999, a week before the earthquake. Although the original terms of reference covered only a revision of the Urban Development Law and its four major Regulations, the work demanded a more extensive survey and revision of the Turkish system of Disasters and Development.

recommendations made in the submitted research report, their source of inspiration.[3]

Apart from the Decrees, most of the steps taken to change the existing state of affairs are done so by means of 'Regulations' and 'Mandates' of the individual Ministries, usually of the Interior, MPWS, or the Treasury. The existing Decrees of the Board of Ministers will eventually be reviewed by the Parliament and have to obtain their consent to become effective as Law in due course, which offers the opportunity for making amendments. There is scope therefore at the moment for every agent and individual for procuring proposals and declaring their opinions for changes.

Most institutions have willingly and often independently contributed to the management of the crisis environment forcing their limits at every possible form of immediate help, the military representing an exceptional standard of performance.[4] All responses of the official and civil administrations or bodies were made with the best of intentions, even if perhaps they still had much to learn from the events confronted. It is crucial that such experience and the unprecedented pool of decisions taken by the Government during the past year, finds its way to build up a new collective system of responding to disasters more effectively. The most valuable outcome will be the institutionalization of the mitigation methods, and the formulation of an overall vision or strategy for sustained administrative conduct.

The scope of this paper is to describe and evaluate the decisions and steps taken by the central and local authorities since 1999, in changing the disaster management system in Turkey into a more rational and effective form, and in so doing the way they are also hopefully converting themselves into parts of a new form of governance despite great resistance and inertia. Such change, in general terms, is a move towards a 'prepared society' which in the context of disaster management, stands opposite to what I tend to describe as the 'fatalist society' (Balamir, 1999). The distinction of the fatalist and prepared societies takes more than crisis and contingency

---

[3] One has to bear in mind that willingness in the adoption of many of the proposals made in our research report by the administration, owes extensively to the impact of the WB report (19844-TU, 1999) to the Turkish Government, whose recommendations run parallel to many of the findings of our report.

[4] Even though the Gölcük Naval Base was itself one of the areas most severely hit, the military maintained security and extensive relief operations, beginning immediately after the tremor. The number of soldiers in the field peaked to 65,000; 114,000 tents were set up, and 45,000 temporary shelters were provided with all the necessary infrastructure and social facilities sooner than all expectations. Hundreds of medical personnel were available in two mobile hospitals and 20 clinics. Schools were constructed to accommodate 1200 students. Together with the expenses for the daily services, the worth of this support amounted to 40 million US Dollars and above.

variants in the style of planning. Rather, these are two separate forms of social existence, and they stand as distinct attitudes towards social organization and conduct of life in total.

This transformation implies institutional and legal changes, the adoption of new tasks and responsibilities by the public and private actors, and the restructuring of the relative functions and positions of existing professions. For this reason, moving from an existing state of relations into another proves to be a painful process. Not only a consensus is rarely achieved in the method of devising proposals for these changes, but often a learning mechanism is hardly available to evaluate the deficiencies, and change the mistakes made. The conventional framework effective since several decades is reviewed below, for an assessment of the recent changes.

2. COMPONENTS AND CONSTRAINTS OF THE CONVENTIONAL DISASTER POLICY

In legal and organizational terms, Development Law and the Disasters Law with their respective attendant regulations constitute the fundamental components of a general policy of disasters in Turkey. Although there are many potential links between the two bodies of law, a first observation reveals the striking fact that the two systems are, as if deliberately, kept apart and alien to each other. The Development Law has almost no reference to natural disasters, whereas the procedures and organs described within the Disasters Law deal fundamentally with the aftermath of disasters. Many of the deficiencies observed in the disasters policy of Turkey can be considered as universal forms of discrepancies, and may make it relevant to have a closer look at these components in their conventional state.

2.1 The Law of Disasters (7269)

The main scope of this Law and its Regulations (1959) is to provide a formal capacity for intervention after disasters, and to organize the relief operations.[5] For this purpose, the Law provides extraordinary powers for provincial and local governors. Such powers are immediately assumed when disasters occur, making the local governor sole authority with powers of commanding all public and private and even military resources and means, property, all vehicles and man-power included. Accordingly, each governor

[5] The actual title of the 'Disaster Law' is lengthy and cumbersome: 'Law Concerning the Precautions and Help to be Maintained Against Disasters Effective on Public Life.' Severn (1995) gives a short history of the preceding regulations.

is responsible for drawing an 'action plan' of relief operations to become effective immediately after a disaster (Severn, 1995, 110). This means almost always however, preparations for 'tents and blankets operations' rather than any form of a risk analysis, estimations of losses and/or a master plan for pre-disaster monitoring of forms of mitigation. These local action plans, as described by the Disasters Law and by the recent mandates of the Ministry of the Interior, are currently prepared with greater attention since 1999.[6]

Of the 68 articles in the main body of the Law, only a few contain provisions concerning the preparations to be undertaken prior to disasters. In practice however, mitigation requirements are hardly fulfilled.[7] In the case of earthquakes, a formal acknowledgement and proclamation is to be made by the MPWS that a 'disaster' has actually occurred.[8] The responsibility for the declaration of different types of disasters like floods, earthquakes, landslides, and fires lies with distinct authorities. If considered a technical necessity by the local authority, buildings could also be identified as sources of high-risk and accordingly dismantled.

---

6 Plans prepared in the town of Gerede (32:12 longitude; 40:48 latitude; 1320m altitude, with a population of 28 thousand, located blind on the unbroken segment of the North Anatolian Fault Line) for instance, cover the formation of hospitals, emergency and relief teams, and task groups such as communications, transportation, infrastructure, damage and value assessment, fire services, traffic and security. As observed in the official files of the local governor, all possible scenarios, procedures and accounting methods are fully described in detail and responsibilities of the tasks are allocated to real persons. According to the plan, an earthquake intensity scenario to correspond Mercalli 9 scale would destroy almost %60 of the buildings in the town and cause loss of life at %0.1 of the population. Official personnel and equipment to be mobilized under the circumstances are 202 persons in addition to which 550 persons (73 medicals, 213 relief personnel, 110 infrastructure experts, 45 emergency operators, 43 social service experts, 14 debris removers, *etc.*) are planned to arrive from the neighbouring provinces. External help also includes 16 ambulances and 111 other heavy work vehicles including lifters, loaders, excavators, fire engines *etc.*

7 An exception is the case of potential landslides, whereby local governments demand the consent and support of the MPWS and Board of Ministers for subsidies. If obtained, a dwelling will be delivered to the household, in exchange with the vulnerable one. The latter is assumedly expropriated and demolished. The determination of a potential hazard in such cases may often turn into a subject of political exploitation.

8 In order to rely on objective criteria, disaster impacts have been reduced to quantitative standards by a regulation, describing the relative level of damage experienced with respect to the size of settlement. According to the regulation, %10 of the population must have been affected in small sized settlements, and at least 50 buildings should have been destroyed in a town of 50 thousand or more buildings, in order to declare a disaster. One third of the agricultural crops lost, is the corresponding standard in the rural environment, which may have tolerances according to seasons and to local capacities and available means of transportation.

Extra-ordinary powers are provided in the Law for the provincial governors to occupy or confiscate land, buildings, equipment, and means of transportation and communication. The MPWS is responsible for recording the damages, vacating damaged buildings and demolishing them, and for the provision of 'temporary accommodation,' as well as 'cash support' from the 'Disasters Fund' as described in this Law. The MPWS immediately revises land-use plans of the disaster-hit settlements. Construction permits are granted, over-riding the existing local plan and municipal decisions if necessary. Settlements hit by a hazard and extensively demolished, or prone to destruction may be relocated in part or in total, according to the decisions of the Board of Ministers, without precedent however in the case of earthquakes.

Valuation of damaged buildings for their expropriation, as well as of land to be acquired for new housing development, are carried out by local committees directed by the local governor, and immediate cadastral services provided. Individuals are entitled to object these decisions through court appeals, but such objections are not to halt the procedures. Owners of damaged buildings or of those property and land reserved for relief development purposes, as well as those owners of property subject to expropriation are designated as 'right-holders.' They are entitled to 'value certificates' equivalent to their losses, which are counter-guaranteed by monies deposited in a specific account. In the case of damaged property awarded by the insurance firms, insurance refunds are deduced from the certificate value of the property. The 'right-holder' households are entitled to have access to publicly developed housing, in conpensation only for a single dwelling unit they own. Construction credits may also be extended to commercial facility owners.

The MPWS determines the location, quantity and other qualifications of housing and other buildings to be produced for relief purposes. Publicly developed properties are thus distributed to the households who encountered property losses in the disaster. These buildings are allocated to the 'right-holders' at their 'cost price.' 'Value certificates' may be temporarily used to meet these costs and to be exchanged with property developed in due course, or alternatively by cash deposited at a specific account. For those 'right-holders' who wish to take the property option, value certificates held will cover part of the cost price of new dwellings. The remainder of debts are distributed over 20-30 years without interest so that housing is practically provided at no cost to the right-holder household.[9]

---

[9] Even if no down-payments (in terms of certificates or in cash) were made, this payment program of debts which according to the Law commences 2 years after the delivery of a dwelling unit, corresponds to one percent of the current construction costs under the current inflation rates of %30-35:

The Law has instituted a Disasters Fund to meet the expenses of operations envisaged, supplemented with annual allocations from the national budget. Transfers to the Fund by the MPWS, and the Ministry of Finance are possible, a prerogative frequently exercised in the case of major disasters. With the centralization of all Funds in a general pool in early 1990's however, the decision for allocations rest currently with the Board of Ministers. Centralization of such prerogatives was not confined to the use of the Funds. Cancelling of debts, changes in the rates of credits, income tax deductions, *etc.*, have been amendments approved by the Parliament (1995). These represent gradual relaxations of technical constraints of the Law and intensification of the populist political trends.

Besides its confinement to post-disaster operations and its content disparate from the Development Law, the Disasters Law and its regulations fall short of constituting a contemporary disaster management system. The Law does not differentiate for instance, between authorized and unauthorized construction, and thereby awards the owners of unauthorized buildings. Building owners who remain strictly within legal boundaries, as well as those who in their rightful decisions obtain insurance for their buildings on the other hand, are punished within the system. Subsidizing the non-conforming members in the society through funds collectively generated by the tax-payers' contributions, remains as an unaccountable mechanism. The law as it is structured and practiced currently, encourages the individuals to ignore land-use plans and choose locations liable to disaster damages, constructing without permits, avoiding insurance coverage, and victimizing themselves so as to receive handsome subsidies. All of this becomes not only a source of injustice but contributes extensively to a culture of fatalism and even leads to the formation of a society in expectation of disasters.

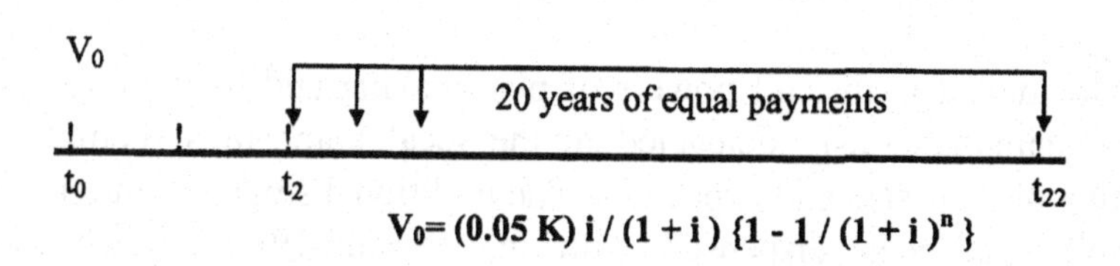

V_0 = total current value of payments
n = number of years
i = rates of interest (%30-35)
K = real total costs of construction

2.2 The Development Law (3194)

The current but outdated Development Law (1985) is the fourth generation in a tradition of such legislation in Turkey. The Law has the declared intention of controlling the 'appropriate formation of settlements and buildings.' Yet this mission of monitoring only the construction stage in development, and aiming to achieve this only in physical terms on singular buildings, is of a very narrow scope. It has little power or incisive tool to manipulate or physically rearrange properties (and to intervene with the sacrosanct rights of ownership), to maintain the 'public good,' particularly to monitor building activity in disaster areas. It ignores the organizational and entrepreneurial stages of development, avoids being involved in the procuring of investments, land assembly, provision of infrastructure and urban services. Furthermore, it is deficient in the technical means of control during the construction stage itself, omits property management approaches, and has a blind eye in the vital need of protection of various categories of (historical, natural, riparian, *etc.*) environment. Thus the Law by contemporary criteria, is outdated and partial in its scope.

As envisaged in the Law, municipal and provincial administrations are obliged to prepare urban plans. In their urban Master Plan-making functions, local authorities are practically free of guidance and of inspection. Yet Master Plans for urban areas represent only an intermediate step in the (currently incomplete) hierarchy of physical plans. The higher and lower level plans and their relation to urban plans are the missing links of the overall system. Urban master plans and procedures for their revisions are the main occupation within the given spectrum of planning activities, regional strategy plans or environmental plans being almost totally omitted. The MPWS, could prepare regional plans if considers necessary, and intervene with the preparation of urban plans where national concerns are involved, or if acute conflicts arise between local authorities. Thus an integrated hierarchical system of plan categories and plans guiding and binding each lower and more detailed ones, and making references to the higher level plans is neither a requirement nor a concern in the current planning practice in Turkey.

A major defect is that all powers of plan-making and their ratification have been simultaneously delegated to the local authorities themselves, irrespective of their size and capacities. The traditional singular authority of the MPWS has been dispensed with in the mid-1980s. Since then, municipalities and provincial governments have been responsible in themselves, from plan-making and from development control functions. Dispersion of such prerogatives has led to great arbitrariness in environmental standards and quality, under the drive for the pillage of rent. The planning system today,with its numerous regulatory mechanisms and actors, is far from a unified body or authority in monitoring physical development.

Since the overall planning control is diffused, it is often difficult to follow rigorously the principles of reducing risks as well. There are almost a dozen of public authorities and ministries other than the MPWS proper, which have assumed through time the rights and powers of plan-making and self-approval under various pretexts. So much so, that it is often enigmatic to settle disputes as to which authority has the ultimate powers at a specific location. This hampers the possibility of a uniformity in the contents and procedures of plan-making, particularly for disaster mitigation purposes.

Besides the multiplicity of plan-making authorities, and the unaccountable and self-ratifying planning and development practices of the local authorities, no effective liability has either been institutionalized in the production of buildings. Building production is carried out with almost no examination of the necessary expertise in projects and no authentic control in the process of production. A system that does not accommodate mechanisms for the inspection of development and building production is bound to generate environments susceptible to hazards. Special is the case of the Ministry of the Interior and the MPWS. The former has vested powers of control over the local authorities, yet has no technical capacity to inspect the plans and projects. The latter on the other hand, is capable of technical inspection yet deprived of its command over the local authorities. Thus the responsibilities of executing the Law (MPWS), enjoying the provisions of the Law (municipalities), and powers of control (Ministry of the Interior) are dismantled and distributed between three parties, which are particularly unrelated and incapacitated.

According to the Development Law, responsibilities of all planning and building supervision are part of the tasks of the municipalities for areas within their jurisdictional borders, and of the provincial governors (in all areas external to the municipal boundaries). They are entitled and responsible for the technical control of projects as well as of their implementation. Even if they intended to, the local authorities are in no position to carry out such control functions, simply for the reason that their financial capacities and their access to technical man-power is considerably shorter than the size of the tasks involved. Building control is therefore almost non-existent in the system, despite the provisions of the Law. In all constructional work, 'building control' is supposedly to be carried out by a 'professional with technical liability,' who in real terms is either hired by the owner or the developer, defying all reliability from the process. As if to complement this, the municipalities are not liable, to any serious extent, for their omission of responsibilities of development and building control. This situation is aggravated by the fact that no legal measures have been taken against officials failing to perform their duties in the spirit of the law.

It is obligatory to obtain building permissions, which could only be given on the basis of projects. Projects submitted to the local authorities are supposed to meet the requirements of both the 'building' and 'disaster' regulations. Building regulations of the Development Law comprise dimensional standards, requirements for heating, lighting, landscaping, parking, fire regulations, *etc.* A few references from several of these regulations are made to the attendant regulation of the Disaster Law concerning structural safety standards in buildings. Only after verifying conformity with the Law and its regulations, that the local authorities are capable of issuing a 'building permission' for any project. This is usually controlled by the municipal officers in terms of physical dimensions, rather than structural safety, the calculations and drawings of the engineer seldom being scrutinized. Following the constructional activity, a 'use-permission' is required before occupation. In this case, site inspection is necessary to establish that the building has appropriately conformed with the permitted project. This again could only be tested in terms of what is observed and measured externally, rather than establishing what remains hidden within the structural qualifications and performance of the building. The intermediate process of construction between the two permissions largely escapes from official inspection of the local authorities to a large extent, even though there are several steps of visa requirements.

The technical liability in the building process on-site is often undertaken by under-qualified individuals who are paid directly by the contractors. Professional inspection services provided by the local Chamber of Architects (not a requirement expressed in the Law), is widely practiced and inheres the same handicap. Such external services could not constitute substitutes for formal inspection proper under liability, since the function is a legal responsibility of the local authorities that could not be delegated. The constraint that buildings will not be allowed to have access to service networks of electricity, water and waste-water discharge systems, unless they obtain use-permission, could be an effective means of control only if it was rigorously followed, and clarification of liabilities of the public agents were to be maintained.

Land assembly and preparation, sub-divisioning, and re-arrangement of property rights are powerful prerogatives of the local administrations but applicable only in areas where rural to urban conversions take place. These powers are the basic operational tools and control procedures of physical planning as envisaged within the 40+ articles of the Law. Disaster mitigation methods in land-use planning and building construction remain entirely external and alien to the conventional system of the Development Law. Avoidance of concerns for disasters in the main body of the Development Law is a conspicuous omission since such concerns can not be confined to the construction of buildings alone. Practice of land-use planning and zoning, transportation and infrastructure planning, procedures for density assignment, planning of open-spaces, participation processes, strengthening and devising of new methods of monitoring building use-control, *etc.* are all distinct aspects of disaster concerns that need naturally be covered in the Development Law. It is strange again that building safety regulations should be accommodated within the Disasters Law. For reasons of practical enforcement and for greater integrity and relevance, the building safety regulation need be annexed to the Development Law *per se*.

Effective implementation of the Development Law has remained such a distant achievement, that separate laws for the control of special areas like 'shores,' 'areas of historical and natural significance,' 'metropolitan areas,' 'tourism centers, 'national parks and reserves,' 'areas of ecological significance' have all been singled out as special cases and covered under distinct laws and separately designated authorities.

Regulations to define the detailed standards of architectural design, installations and mechanical network linings within buildings, guiding principles for interior furnishing, street furniture and building exterior surfaces *etc.* are completely ignored in the system of disaster management. Apart from the powers of *eminent domain* and that of imposing shared-easements (better known as the article 18 applicable only at urban fringes subject to development for the first time), there are no effective tools in the planning repertoire. Much is necessary however, in a system of disaster management, efficiently to protect areas subject to disasters from development, to improve the environmental standards and resistance capacities of existing built-up areas, to avoid disaster chains by strict land-use control, *etc.*, all requiring greater efficiency and more powerful tools. Disaster management is one such area of activity that a case for more intensive planning control powers could have utmost legitimacy.

Since specially standardized geological maps and maps, as well as integrated information related to other disasters are not considered a prerequisite in the development system, an objective basis for the evaluation of land-use and building permission decisions for their contributions to mitigation efforts does not exist. Geological evaluation reports for individual sites as required by some municipalities, are piecemeal and can not be impartial because they are prepared by the investing party. In cases where more comprehensive geological/seismic *etc.* information and recommendations are available, no formal method of taking these into account in the practices of land-use planning exists.[10]

---

[10] The case of Gerede is not probably the only extreme example of how all physical development could be shaped in the most undesirable configuration, despite all warnings and recommendations. The 1944 earthquake in Gerede took place on the 1st of February,

3. RECENT STEPS TAKEN FOR DISASTER MITIGATION: IS TURKEY MOVING TOWARDS A NEW DISASTER STRATEGY?

Structural changes were made in the legal and organizational framework concerning disasters in Turkey, by means of the Governmental Decrees produced since 1999. These are based on the Law 4452 (27.8.1999) authorizing the government in its actions for the necessary immediate and long-term measures concerning disasters. Unlike the tendencies and behaviour symptomatic of post-disaster periods, that aim only to extend the scope of distribution of resources and rights to the victims and their relatives, a number of decisions were taken and enforced by the government through these Decrees which could be interpreted in no other terms but an indication of a new attitude in structuring disaster policy.

The introduction of the institutions of 'obligatory earthquake insurance,' 'construction inspection' functions, and provisions for the improvements in 'professional competence' have been three inducive decisions taken by the Government that implicitly restructure the formal response attitudes to earthquakes in Turkey, clearly indicating an unusual preference for the mitigation alternative. In organizational terms as well, several attempts aimed to accomplish a more comprehensive management system. Apart from extensions made in the responsibilities of the local authorities in disaster mitigation, three complementary organizations were introduced. Ministry of the Interior has set up regional centers for relief and emergency operations. A General Directorate of Emergency Management established and attached to the Prime Ministry, and an independent National Earthquake Council formed by a Prime Ministry mandate were the two further moves of the Government in the fulfilment of functions hitherto ignored. These new

---

6:21 in the early morning for seven seconds. Witnesses claimed underground tremor from the north, an initial movement from east to west then reversed; also a rotational push from bottom. Some have also observed lighting at the atmosphere. According to the Geological Report of 1944 most of the damage occurred along the East-West main fault-line. Irrespective of its material and make, all buildings standing on this crack extending for kilometers at both sides, have been destroyed. The northern side of the crack has at places has collapsed down 80-100cm. Secondary cracks not exceeding 100-150m have also appeared. Almost 709 buildings were seriously damaged out of total 901. 69 persons died in the town. Recommendations were that: (1) No buildings should be erected within 50 meters of the main fault line; (2) No construction should be allowed within 25 meters of the secondary cracks; (3) Sandy soils, as in the case of Demirciler quarters should be avoided; (4) Safer areas for settlement are the hills south of the motorway. The current developments in the town totally deny these explicit requirements. The geological report of 1944 (access for which demanded archaeological digging) never found its way to enforcement.

provisions deserve a closer review for their potential contributions in transforming the disasters strategy in Turkey.

Figure 1: Conventional System and New Provisions in Disasters Policy

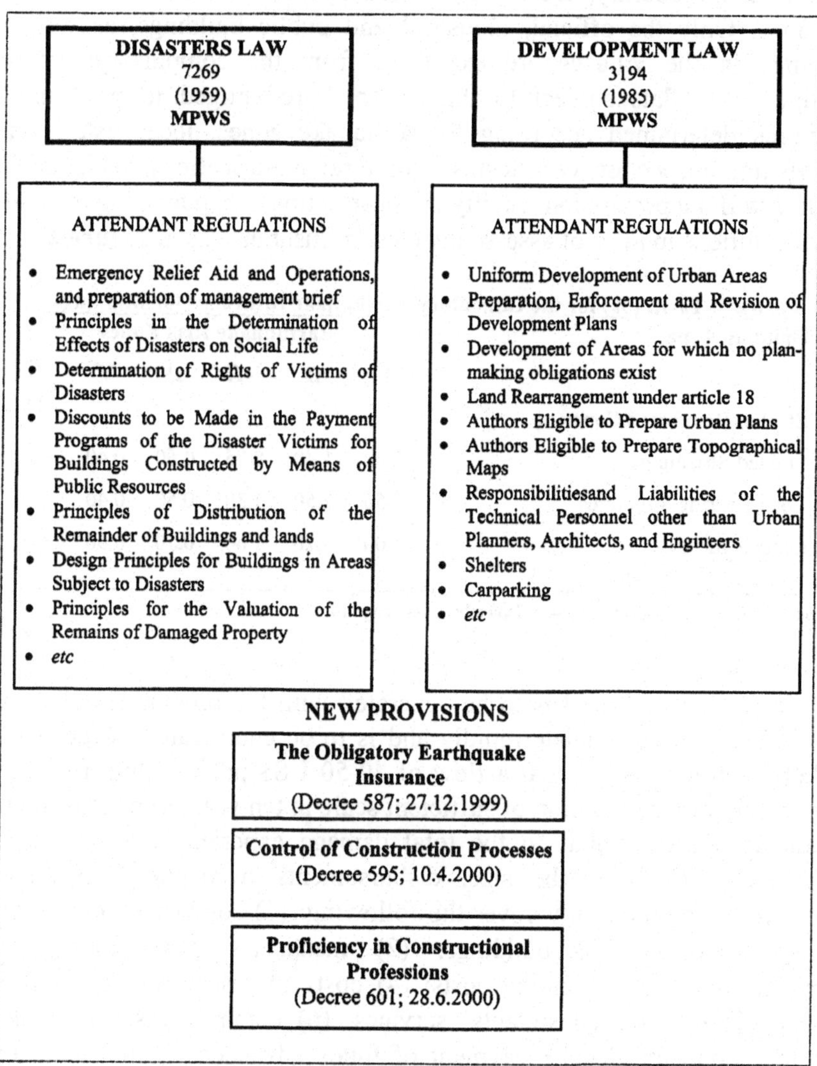

3.1 The Obligatory Earthquake Insurance (Decree 587; 27.12.1999)

A Natural Disasters Insurance Administration is established attached to the Treasury. The Statutory duty as determined by the Disaster Law for the compensation of losses of disaster victims is terminated with this Decree, and the Natural Disasters Insurance Administration is taking over the responsibility. Beginning in year 2000, all residential buildings and independent sub-units of buildings registered at the cadastral records throughout the country, are to be covered by the compulsory earthquake insurance. Only the official, industrial and public buildings, as well as buildings in the villages are exempted from the compulsory system. Buildings and flats subject to this scheme are obliged to pay annual premiums determined according to earthquake zones, local risk levels, construction inspection certificates, structural modifications made in the building without permission, quality of construction, *etc.* punishing the more risky conditions in rates of assessed values for insurance as in **Figure 2**.

Figure 2: Tariff (%) for the Obligatory Earthquake Insurance for Buildings

Building Type	Earthquake Risk Zones				
	I	II	III	IV	V
Steel or Reinforced Concrete					
Framed Structures	2.00	1.40	0.75	0.50	0.40
Load-Bearing Structures	3.50	2.50	1.30	0.50	0.40
Other Structures	5.00	3.20	1.60	0.70	0.50

Source: Official Gazette no.: 24164, 8 September 2000, page 14.

A large financial pool is to be generated with the premiums collected. This is likely to accumulate rapidly and is to be enumerated in the world markets. Estimations are that a flow of 40-50 US$ per dwelling per year, could reach an accumulation of sufficient scale in ten years so as to refund a volume of losses similar to the total damage experienced in the 1999 earthquakes. Whatever the size, disbursements from the Compulsory Insurance Fund could only cover the following: (1) compensations for the damaged and eligible buildings; (2) manager's commissions; (3) Administration's own running costs; (4) costs of research and scientific studies; (5) fees for consultants' services; (6) commissions for eligible insurance companies; (7) repayment of funds advanced by the State; (8) costs of damage assessment services; (9) public relations and campaigns.

The Administration is to tender the management of the Insurance Fund to firms, for periods of 5 years. The refundable losses are to be determined by the Administration within the limits set by the Government. For the year 2001 the maximum refundable damage is not to exceed 20 billion TL (less than 30,000 US$). The Administration could employ consultants (national and international) for research and development on contract basis. The

Administration itself is directed by an Executive Board of seven persons: 4 high level public officers, 2 high level insurance and finance organization directors, and an earth scientist.

A voluntary dismissal of the prerogatives of spending public resources for political purposes and surrendering of such privileges to a relatively autonomous insurance administration in itself is nothing short than a heroic move on behalf of the current government. The promise that only property covered by the insurance will be eligible to receive compensation is a revolutionary idea within the traditional practice and realities in Turkey. It still remains to be proved however that the populist trends could be resisted and the political bodies restrain themselves provide donations to the owners of unauthorized buildings in the face a disaster. In this vein, provisions of the Disasters Law that oblidge governments to give aid and accommodation to all victim households will have to be disposed of.

Figure 3: Priorities in Risk Management

1
RISK AVOIDANCE
Avoidance (avoiding areas of natural risks for settlement purposes)
Distancing (specifying minimum distances from nodes of hazard)
Refusal (prohibiting combinations of uses to cause chain-disasters)
PLANNING SERVICES
plan preparation/ plan implementation/ plan control/ plan revision services
2
RISK MINIMIZATION
Discarding Risks At Source (*eg.* flood-control reservoirs, artificial avalanches)
Upgrading Resistance At Location of Effect (*eg.* levees, building codes)
ARCHITECTURAL AND ENGINEERING SERVICES
higher standards in disaster-resistant design and construction/
inspection of construction/ building retrofitting services
3
RISK SHARING
Aids and Subsidies (credits, rent subsidies for dwellings and business premises)
Donations (national/ international, voluntary/ organized, cash/ in kind donations)
Taxes (extra burdens on others than those suffered losses in the disaster)
Insurance (compulsory building insurance)
FINANCIAL MANAGEMENT SERVICES
building up funds for pre-disaster and post-disaster requirements/ efficient allocation of funds/ maintaining equity between fund-providers and between fund-users

A second admirable aspect of this move is the formation of a very large pool that is independent of the national budget and therefore to a large extent free of the molest of political decision makers. Payments made to an independent body operating for common good in the long term is far more preferable for the citizens, than the burden of a short-sighted governmental levying of extra taxes. This is particularly true after the double-taxation of all property and vehicles, imposed to every citizen during the past year for funding the rehabilitation programs.

These two salutary aspects aside, the insurance system in effect has appalling deficiencies. The gravest and most obvious one is the unnecessary obstruction of flow of funds to mitigation investments. Many reasons may be advanced why a modest part of the annual incomes of the Insurance Fund should be dedicated to risk avoidance and minimization projects, revisions in land-use planning, and retrofitting efforts in public and private buildings. In the first place, the basic principles of risk management, as currently adopted by the World Bank (Kreimer, *et. al.*, 1999) and other authors (Burby, 1999), demand that risk avoidance and risk minimization provisions should have priority over risk sharing (**Figure 3**).

Irrespective of the size of compensation available for potential losses, saving of lives, property and other assets with reasonable levels of investments stand as the rational alternative. When human life is involved, the efficiency of mitigation investments is bound to remain high. A partial allocation (*eg.* %20) of the annual incomes of the Insurance Fund to mitigation investments would make a perfect excuse against likely resentments to this tax-like payment of premiums. Improvements in city plans, infrastructure and retrofitting of public buildings as programs of preparation for disasters are to contribute to the tolerance and willingness of people in participating the insurance schemes, as against the alternative of maintaining huge funds at great costs only for compensations after the disasters. On the other hand, it is most unlikely that the Turkish society at large could financially and politically afford generate another fund of similar magnitude for the realization of mitigation programmes. The property owners and households could not be approached once again for raising a separate fund to carry out mitigation work. A trade-off is therefore inevitable between compensation and mitigation funding for the legitimacy of the operations of the Insurance Fund in administrative, social, and psychological terms.

A decision of equal gravity is the exclusion of technical professionals related to mitigation work (like urban planners, engineers, architects, *etc.*) from the Executive Board of the Insurance Administration. The system is likely to recruit technical professionals for their services of inspection, assessment, as well as for independent evaluation services in courts of many of the potential disagreements dormant in the relations formally described. Employment of the technical professionals by the finance sector is likely to give rise to inverse relations in the sense that planning and design functions will become subservient to the puposes and aspirations of the finance sector. The macro social objectives are not to be structured for maintaining high premiums at the expense of poor development, but secure environments and buildings of sufficient safety. The priorities here are in the improvement of physical planning and control systems. This could partially be achieved if plan-upgrading and building-retrofitting projects could be subordinated by the Insurance Fund through regular allocations of resources for competing projects submitted by the local authorities and other bodies.

3.2 Control of Construction Processes (Decree 595; 10.4.2000)

Vital variances observed in the performance of buildings after the recent earthquakes have convinced everyone that building production should be held under closer inspection and responsibilities of parties involved be unchallengeably clarified. With the consensus that local authorities have always fallen short of controlling constructional activity, the tasks and responsibilities of inspection were entrusted with special firms to be instituted for the purposes, keeping meanwhile the right of ratification with the public authorities as a Constitutional requirement. The general idea of structuring a private service under official scrutiny, as well as many of the details of an inspection system were directly taken from proposals of the research report submitted a week before the earthquake.

The Decree introduced private Building Inspection Firms (BIFs) entitled to control all projects and constructional activity and report to the local authority responsible for the permissions of construction and that of occupation of buildings. These firms could only be instituted if a minimum %51 of their capital assests belong to eligible architects and engineers. BIFs are obliged to control projects, all building activities, standards of materials used, and geotechnical reports. They are also to keep records of progress and submit their reports to the local authority providing the permissions. There are three categories of such firms entitled according to the size of establishment, composition of personnel, and eligible to inspect projects and building activities of different scales. Only upon the positive reporting of a BIF, the local authority is to ratify a project and issue the construction or occupation permit. BIFs are to operate under financial liability insurance. All buildings other than public, and those larger than 180m^2 are obliged to employ BIF services. This is likely to increase the overall costs of these buildings by %4-8. Thus fees of BIFs are also regulated. Buildings constructed under inspection are registered and given certificates. Inspection information plates are permanently exhibited on these buildings.

Supervision of BIFs are organized at three levels. At the provincial centers and sub-centers (all settlements of 50,000 population and above) Inspection Committees operate under the coordination of local director of public works and settlement, and are formed by the representatives of the municipality and the related chambers of professions (architects, city planners, engineers). In the smaller settlements, BIFs are kept under surveillance by the Committees of the provincial center. The Committees follow the activities of the BIFs, keep their performance records and of the parties involved (developers, construction managers, authors). They are also to resolve disagreements between them. The High Committee operate at the Ministerial level granting licences to BIFs, keeping their performance records, giving penalties or taking their titles back. This central Committee is composed of representatives of the Ministry of the Interior, Insurance Directorate of the Treasury, Union of Chambers of Engineers and Architects, Union of Chambers of Commerce and Industry.

The Decree has clarified the responsibilities of the parties involved in construction (*ie.* property owner, developer, construction manager, author), and standardized the agreements to be drawn between them. A special and protected account in the name of each local authority is to be operated. The property owners will deposit the amount to meet the services of BIF in this account only to be transferred to the BIF with the consent of the owner and approval of the local authority. Inspection services could only be carried out by those professionals (architects and engineers) affiliated with a BIF whose expertise are recognised by the related professional chambers. In granting a construction permit, the local authority will demand the BIF licence, liability insurance documents, and the standard agreements between the parties involved, apart from those required by the Development Law.

Constructional activities will be stopped at any site, if a related BIF is to resign and no replacement is achieved by the owner within 20 days. BIFs are not allowed to undertake inspection work involving buildings owned by relatives or business partners of the professionals. In return, professionals will not take responsibilities in another BIF, neither will they be allowed to work in the capacity of a developer or manager in any construction. BIFs are directly responsible of all constructional defects, damages experienced due to expected disasters, and are liable to immediate payment of compensation for the damages. Such compensation could afterwards be receded to other faulty parties by the BIF. BIFs are liable for ten years in the case of structural damages, and for two years in other types of defects. BIFs will not be responsible however, of the changes made by the owner in the original permitted design, and in issues that have been subject to previous warnings.

One of the most commendable side-effects of the Decree that goes unnoticed is its bridging of the local authorities in their local context. The provincial governorates and municipalities in the conventional model, have an incongruous relation to each other in their respective responsibilities and liabilities concerning building activities and disasters. Although the provincial administrations are fully responsible and liable for all activities and losses in buildings after a disaster, they have no powers to control or intervene in building processes in ordinary times. Municipalities on the other hand, ordinarily have all powers of monitoring planning and constructional activities. They can enjoy therefore the political freedom and 'irresponsibility' in the carrying out of this function, reluctant controllers of development as they are, since they are almost beyond reproach in cases of liability accusations. The formation of joint Inspection Committees by this Decree is a first ever attempt in bringing together the two administrations that have long been standing apart, in tackling the problems of 'shared risk,' removing the obstructions to acquire capacities on the way of evolution towards superior forms of 'adaptive systems' (Comfort, 1999).

The provisions of the Decree, which for the moment are operable in 27 provinces, bring a formal and real construction inspection mechanism for the first time in Turkey. This is a major contribution to mitigation efforts covering the construction of new buildings designed *ab initio*, and thereby generating a segmentation in the property markets. The stock inherited however, remains as an untouched problem area. In terms of earthquake risks, production of safer buildings may still have little significance in certain areas where new buildings are constructed in old neighbourhoods and where new buildings are equally susceptible to environmental damages to be incurred. Without appropriate means and tools of land-use planning therefore, that take into account seismic risks, individual building safety may have little meaning. After all, the principles of risk management give higher priority to the avoidance of risks than risk minimization efforts.

There are other contradictions and deficiencies of the Decree which are expected to be revised during the procedures leading it into the status of a Law. The possibility of public bodies for instance, establishing their own inspection units, has been dismissed. Secondly, the determination of a BIF by the free choice of the property owner could be curbed with the designation of the firm by the local authority and/or the professional Chambers. Thirdly, structural safety concerns have brought the engineer into fore-ground and promoted the role of engineering tasks in the construction sector. This over-emphasis tends to challenge the relative status of the related professions and undermine the conventional superior rights and legal responsibilities provided by other laws of the designer-architect in the orchestration of the building activity.

3.3 Proficiency in Constructional Professions (Decree 601; 28.6.2000)

This Decree changed a number of points in the existing 'Law concerning Engineering and Architecture' and the 'Law of Union of Chambers of Engineers and Architects.' The amendments describe the requirements for improved professional competence in the fields of engineering and architecture. A minimum of five years of professional practice, attendance in the training courses and success in written examinations organized both by the related Chambers are the essential conditions. Persons in demand of services of engineers and architects may require qualified professionals in the production of safer buildings to resist hazards and earthquakes.

According to the Decree, the qualifications identified are necessary only for those services that require competence. These services are not however clarified in the Decree itself, but determined elsewhere, as in the case of Decree concerning Building Inspection Firms. The non-obligatory terms used for the description of employment of competent professionals may be considered a major weakness. The provisions have significant implications however, on professional performance in general and high potential in the improvement of professional education.

3.4 Organizational Moves

Organizational rehabilitation and establishment of new and complementary units was inevitably taken into the agenda of the government during the past year. Earthquakes of 1999 gave great impetus to the existing organizations, in their reviewing of capabilities, and devising more efficient methods of work. Besides (the General Directory of Disasters of the MPWS, and the Observatory attached to the Prime Ministry and operating as a branch of the Bosphorus University), the two existing official institutions directly related with earthquakes, several steps were taken. In the first place, responsibilities of the local authorities were extended by Governmental Decrees to cover disaster mitigation efforts by amendments to existing Laws. The other moves for organizational reform have been the following:

3.5 Directorates of Civil Defense for Rescue and Emergency Attached to the Ministry of the Interior (Decree 586 and 596; 27.12.1999 and 28.4.2000)

The Ministry of the Interior decided to set up 11 provincial directorates equipped with necessary vehicles and devices, recruiting 2500 permanent persons and 300 on contractual basis, and to prepare detailed local plans for their activities, training and occasional drills. These are to reinforce the provincial 'rescue and aid committees' and local relief forces with more professional and alert reserves at strategically stationed regional centers.

3.6 General Directory of Emergency Management attached to the Prime Ministry (Decree 583; 22.11.1999)

The unit first established with a staff of 16 persons as a directorate was soon after promoted to a General Directorate responsible for high level coordination. The functions of the General Directorate are as follows: (1) coordination of post-disaster activities; (2) formation of emergency management units in public organizations; (3) taking of disaster mitigation measures; (4) short-term and long-term planning of related tasks; (5) formation and management of data-banks; (6) coordination of relief equipment and motor vehicles; (7) formation of scientific, technical, administrative committees.

3.7 The independent National Earthquake Council (Prime Ministry Mandate 2000/9; 21.3.2000)

This Council of 20 scientists has been instituted by the Prime Ministry, owing to the chaotic environment created by the contradictory claims of earth scientists in particular and exploited by the media industry during the past year. An authority was considered necessary which could make the final assessment of events in relation to earthquakes, and point to the necessary lines of action. Members were identified by universities and related institutions, each nominating individuals other than their own. The 20 scientists of the Council are distributed according to the related disciplines of 8 earth scientists, 8 structural and earthquake engineers, 4 other fields (currently composed of an architect, a planner, a social psychologist, and an environmental engineer). The tasks of the Council are identified as: (1) scientific assessment of earthquake predictions and informing the public; (2) identification of priority research areas concerning mitigation; (3) consultancy to public bodies and the development of policy and strategies; (4) ethical matters concerning earthquake prediction.

The Council has determined its *modus operandi,* made several announcements, and at the moment developing a report on National Earthquake Strategy.

4. THE INCOMPLETE TASK: AN EVALUATION OF THE CURRENT STATE

It was a most unfortunate coincidence that immediately after our completion and submission of an extensive research work to the MPWS, the events of 1999 took place. The research explored the possibility and viability of incorporating disaster mitigation methods within the development system. It was not gratifying either that many of the recommendations of this research were put into effect under the compelling circumstances of the day. To evaluate the evolving disaster mitigation system in Turkey, its current state, and to establish what remains yet to be achieved, it may be relevant to compare this progress with the propositions of the research.

1. The need for a settlement-based uniform geological information system still remains unattended. This task of preparing 'Coordinated Disaster Maps' had a first priority among the recommendations of the research project. Accordingly, the General Directorate of Disaster Affairs of the MPWS is to upgrade its functions of preparing geological maps and render services under the directives of the proposed High Council for Disasters. The financial resources for the task is to be met by the local authorities involved and by the Disasters Fund (which could now be redefined as the Earthquake Insurance Fund). Nothing has been officially provided at the moment on this issue, apart from mandates and instructions of the Ministry to local authorities to the effect that for planning revisions should be made to include large scale assessments of the geological conditions for areas in their jurisdiction. This implies distinct efforts of local authorities glean resources to hire private geological firms, the services of which are not described in terms of scope, standards or method.

2. With the introduction of building control and insurance functions, and professional proficiency measures in the Decrees, a decisive turn has been achieved on pre-disaster considerations and preparations. For the first time, the post-disaster bias has been overcome and attention is focused on the tasks of mitigation to be carried out prior to a disaster. Yet the primary subject of development planning and control is omitted. The provisions refer to the future formation of singular buildings rather than cities in their collectivity and totality, and exclude the problems of existing and inherited stock of buildings, let alone the unauthorized forms of development. In the recommendations of the research, operations in high-risk areas are routinized in long-term programs within ordinary practices of land-use planning. Evaluation of 'Coordinated Disaster Maps' and identification of 'high risk areas' by the planner is a first exercise in the generation of 'contingency plans.' Designation of 'Disaster Action Areas,' preparation of detailed application plans and 'Disaster

Action Programs' are the other implementative tools of such planning.

3. Distancing decisions related to disasters from political concerns and devising decision support systems is a fundamental problem of disaster policy. In Turkey, the trade-off between political and technical decision-making has considerably changed with the recent moves. The powers of distributing resources have largely been surrendered to the Building Insurance system which will have to rely on objective criteria in its operations. Building inspection procedures and proficiency measures are likely to promote the technical labour and know-how, further curbing the politically manoeuvrable area. Although the recommendations of the research found direct responses in these points, other suggestions to distance political activity from disaster mitigation efforts such as the establishment of a High Council for Development, and a High Council for Disasters have not materialized, even if the National Earthquake Council partly meets the latter.

4. Changes in the nature of responses to disasters, from spontaneous and extra-ordinary decisions taken immediately after shocks of disasters, to routinized procedures of decision-making for mitigation is another criterion to judge the state of a disaster policy system. A move in this direction was again achieved. Insurance of buildings, or the inspection activities are standard and routine actions spread over time. Procedures introduced by the Decrees will again serve to reduce the need for taking extra-ordinary decisions after each earthquake simply for the fact that more intensive pre-disaster mitigation efforts will imply less of losses and damage. Therefore the more mitigation processes are instituted, the more routinized the whole system will become.

5. The setting up of the Insurance Fund is an achievement that cannot be exaggerated. In terms of its sources, pace of accumulation and size, it will prove of global significance. Yet the current scope envisaged for its use is too narrow. Any Fund allocated for disasters has to encourage and facilitate mitigation work rather than be kept solely as a reserve for compensations. No compensation for human life could be better justified than the taking of all possible means of precaution to protect life and reduce it to minimum possible levels in the first place, rather than tolerating loss of life with the consolation of keeping immense resources at hand to be paid for its assumed worth.

6. Rationality in risk management requires a loyalty to principles of priorities (**Figure 3**). Provisions of the Decrees stand in contradiction to all rational approaches. Methods of 'risk avoidance' have been ignored and efforts were spent to devise methods of 'risk sharing' (Compulsory Building Insurance and related Fund), and the 'minimization of risk' measures (Building Inspection and Professional Proficiency systems). The most effective and relevant method of risk avoidance system is to be maintained by land-use planning proper informed and supported by the earth sciences. This has been the terms of reference of the research undertaken. Structuring of an official and binding system of formal disaster maps for settlements is a pre-requisite for land-use planning, a primary step for any mitigation program.

Land-use planning according to Disasters Law is considered as a 'remedial' post-disaster operation. There are other approaches that consider the post-disaster environment as an 'opportunity' to 'improve traffic, eliminate non-conforming uses, modernize public facilities, and stimulate the local economy' (Spangle, *et. al.*, 1991). According to the approach subscribed in the research however, land-use planning is a device for risk avoidance and minimization, which is not necessarily a new innovation since others have previously attended the same line of thinking (Germen, 1980). The procedural requirements and contents of this alternative however, implies distictly different agendas compared to existing systems of land-use planning (**Figure 4**).

Figure 4: Land-Use Planning in Settlements of High Risk

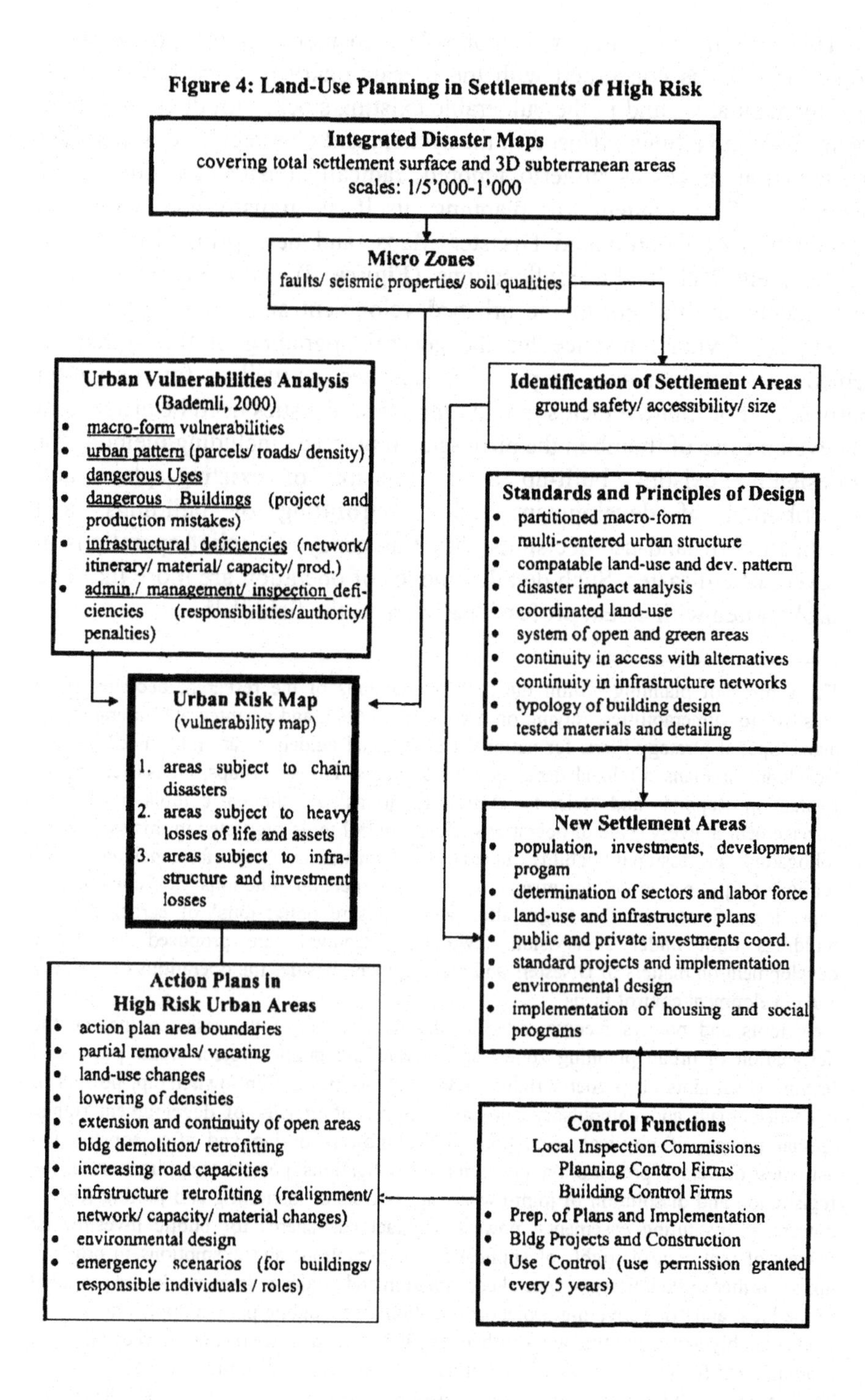

This attitude in planning is not solely focussed in the post-disaster conditions, but is concerned with the precautionary practice both in new development areas and in the vulnerable existing stock. Introduction of new agents and obligations diffused in the system, and new regulations enabling participation processes hitherto ignored, radically change the context of planning.[11] The content of planning itself is transformed with the introduction of Coordinated Disaster Maps, and new principles for the development and land-use allocations (**Figure 4**). Introduction of new instruments and powers to the urban development and land-use planning system is of vital relevance for the general upgrading of the system to remedy the consequences of past four decades, as well as for mitigation purposes.[12] For improvements in planning, several issues need be articulated to develop rules of thumb in the professional practice, including methods for re-designing existing built-up areas, revision of existing plans and redistribution of development rights, retrofitting of buildings, and coordination of land-use decisions with emergency and relief operations in the event of a disaster. Such detailed modes of operation are more likely to be proliferated with actual professional work in the field.[13, 14]

---

[11] The context of planning within the recommendations of the research, becomes more sensitive to vulnerabilities, relying on a complete system and an integrated hierarchy of plans. Spatial Strategy Plans for national and regional guidelines are introduced besides Development Plans for local decisions. 'Coordinated Disaster Maps,' 'Disaster Impact Evaluation Reports' and 'Disaster Action Programs' are the basic tools of disaster management at every scale of operations. Other control tools suggested in this system are 'obligatory professional liability insurance,' 'constraints in professional conduct,' 'obligatory protection of documents related to inspection functions' for construction supervision firms, and the 'registration and exposure obligations' of certificates for buildings built under supervision. Two High Councils are proposed for policy development on matters of Disasters and Development, besides the operations of building and development control firms.

[12] New tools and powers recommended in the final report of the research include the designation of urban planning zones, and private and public project areas/ powers of levying constraints on property rights/ powers of instituting joint-ownership, powers of eminent-domain and compulsory purchase/ powers of transfer of development rights/ special powers of physical regulation/ control powers during and after construction activities/ powers of pre-emption and reciprocal obligations/ powers of application for the registration and dissolution of joint-ownership/ powers of instituting and joining REITs/ powers of instituting easements/ powers to exact cost-shares for public investments/ powers of rent control/ rights of application for exceptions and exemptions in property taxes. Further to such powers, special coordination and guidance frameworks are provided to the local authories covering project application areas/ public project application areas/ land assembly and regulation areas/ urban renewal areas/ disaster risk areas/ receiving and sending areas for TDR/ special use and building areas/ areas for public property.

[13] These points have been taken up by two separate academic exercises at METU during the 1999-2000 terms, for two settlements subject to earthquake risk. In the case of

Based on a new and broder framework that empowers urban development and land-use planning, disaster mitigation efforts are likely to flourish finding their way to generate tactics with reference to local conditions, and at times introducing unique methods appropriate and relevant for the case at hand, contributing to the global knowledge and experience.

---

metropolitan city of Bursa, (Bademli, 2000) apart from an 'Integrated Disaster Map' which provided the measures for different local exposure intensities to seismic hazard obtained from earth science experts, an 'Urban Faults Map' was developed through sampling methods (**Figure 4**). These components were then combined in a final 'Urban Faults Map' and overlaid to the 'Integrated Disaster Map' to obtain the 'Potential Damage Distribution' in the Metropolitan area.

The study of Gerede, a much smaller settlement, on the other hand, represented a very different context and combinations of risk conditions (Balamir, 2000). Since the whole settlement is aligned on the very global fault line, every effort to halt development in the existing area and develop attractive means of relocating the industry, the commercial center, and the public buildings to relatively safer areas even if at considerable distances, and condemning most of the existing settlement seemed to be the only physical solution from the very beginning. More of the efforts in this case were therefore concentrated on the administrative and social preparedness in the short term, rather than be confined to measuring the physical differential risk distributions alone. Risk reduction recommendations were developed based on the surveys and analyses of the public infrastructure in detail. These covered the layouts of water, drainage, electricity and telephone networks and their superimposition with the system of major and minor faults, calculating the crossings over fault lines. Based on unit costs of each network, areas and population served by the branches, risk levels at point locations can be determined, then the costs of operations of religning could be compared with values at stake in case of an expected earthquake damage. For the monitoring of urban development, trends in urban population growth, investment capacities in building construction, and costs of development as compared to existing stock values of highest vulnerability are used as comparative indices for relocation decisions. Estimations for extra-funding of such operations are then made.

[14] Professor Bademli works also as a consultant for the Gretaer Municipality of Bursa, and many of the ideas and findings of this research, particularly those related to the unauthorized settlements within the Bursa Metropolitan area, are in the process of trial in real world within the 'Healthy Cities' programme and 'Agenda 21' activities.

REFERENCES

Bademli, Raci. 2000. *Disaster-Safe City: Bursa*, research reports of the City Planning Studio Work of the fourth year Undergraduate Program directed by R. Bademli, in the City and Regional Planning Department, METU, in preparation for publication, Faculty of Architecture, METU.

Balamir, Murat. 1999. Reproducing the Fatalist Society: An Evaluation of the Disasters and Development Laws and Regulations in Turkey, *Urban Settlements and Natural Disasters*, ed. E. Komut, Chamber of Architects of Turkey, 96-107.

Balamir, Murat. 2000. *Enabling Local Administrations and Communities to Cope with the Earthquake Hazard*, research report of the City Planning Studio Work of the Postgraduate Program directed by M. Balamir, in the City and Regional Planning Department, METU, in preparation for publication, Faculty of Architecture, METU.

Burby, J. R. 1999. Unleashing the Power of Planning to Create Disaster-Resistant Communities, *APA Journal* (Summer) 249-258.

Comfort, Louise K. 1999. *Shared Risk: Complex Systems in Seismic Response*, Pergamon, Elsevier Science Ltd., Oxford, UK.

Germen, Aydýn. 1980. Town Planning as a Response to Earthquake, *Proceedings of the 7th World Conference on Earthquake Engineering*, 8-13 September (volume 9) 277-284.

Gülkan, Polat, Balamir, Murat, and Sucuoğlu, Haluk, et. al. 1999. *The Revision of the Urban Planning and Building Control System in Turkey for the Mitigation of Disaster Effects*, final report of research prepared for and submitted to of the Ministry of Public Works and Settlement, supported by the World Bank, Ankara.

Kreimer, Alcira, et. al. 1999. Managing Disaster Risk in Mexico: Market Incentives for Mitigation Investment, The World Bank, Washington D.C.

Severn, R. T. 1995. Disaster Preparedness in Turkey and Recent Earthquakes, in Erzincan, in D. Key editor, '*Structures to Withstand Disaster,*' The Institution of Civil Engineers, Thomas Telford, London.

Spangle, W. and Associates Inc. 1991. *Rebuilding After Earthquakes: Lessons from Planners*, International Symposium on Rebuilding After Earthquakes, Stanford Univrsity, Stanford, CA, August 12-15 1990, California.

INSURANCE AND RISK TRANSFER MECHANISMS

12

Alternative Risk Transfer Mechanisms for Seismic Risks*

Martin Nell
Johann Wolfgang Goethe-University
Frankfurt/Main, Germany

Andreas Richter
Hamburg University
Hamburg, Germany

1. INTRODUCTION

Damages inflicted by natural catastrophes in recent years have accounted for economic losses of a size heretofore unknown.[1] During this period, one could detect an increasing frequency of catastrophic events as well as an increase in the average amount of loss per event; the latter largely stemming from the geographic concentration of values in catastrophe-prone areas. For the case of earthquakes no significant trends in the number of occurrences are observed, but the influence of concentration of values on damages was demonstrated in a dramatic way in 1999: Although the number of severe earthquakes was not unusual, these events, among them the dreadful disaster in Izmit (Turkey), were

* The authors would like to thank Klaus Bender, Petra Lenz, Nadia Masri, Dirk Sanne, and Winfried Schott for helpful comments. Of course all errors and omissions remain the responsibility of the authors. Project support from an academic fellowship granted by the German Academic Exchange Service (DAAD) under the Gemeinsames Hochschulsonderprogramm III von Bund und Ländern (Richter) is gratefully acknowledged.

[1] See e.g. Cummins/Lewis/Phillips (1999).

P.R. Kleindorfer and M.R. Sertel (eds.), Mitigation and Financing of Seismic Risks, 237–253.

perceived as a very singular accumulation, since in a short time span several densely populated areas were hit.[2]

As examples of further extreme catastrophes, reference is usually made to earthquake hazards in Tokyo or California. The estimated loss potential (PML) of these risks seemingly shows the capacity limits of traditional insurance markets. For instance, estimations of insured losses after a major earthquake in the San Francisco area amount to approximately $100 billion; on the other hand, balance sheets of the U.S. property liability insurance industry show a cumulative surplus of about $300 billion.[3]

These "capacity gaps" in the industry[4] have been at the heart of many discussions among insurance economists and practitioners in the recent past, largely aimed at the development of possible solution strategies involving the financial markets, which are usually referred to as Alternative Risk Transfer (ART). Contributions can be expected, if, for example, the issuance of marketable securities was able to attract additional capacity from investors who are not otherwise related to the insurance industry. In practice, rudiments of this kind can be observed in various forms since 1992, even though they have yet to reach a significant market share.[5]

This paper is concerned with a certain kind of ART instrument, namely catastrophe index-linked securities (indexed cat bonds). A cat bond is a contract between an issuer and an investor. The investor puts up an amount of cash at the beginning of the coverage period; this is held in escrow until either a catastrophe occurs or the coverage period ends. The issuer offers a certain coupon payment, provided that no catastrophe occurs, at the end of the period and returns both principal and interest to investors. In the event of a catastrophe, the investors will receive no coupon payment, and some or all of their principal may go to the issuer.[6]

It has been shown by Croson and Richter (1999), that contingent capital, and particularly indexed cat bonds, would be a very useful instrument for a sovereign government. It could help to avoid disruption of important infrastructure projects, caused by the fact that in case of a catastrophe all liquid resources might be devoted to humanitarian aid. The possibility of either sudden financial distress due to a catastrophe or the adoption of inferior construction strategies because of the potential for

---

[2] See Swiss Re (2000), pp. 10.

[3] See e.g. Cholnoky/Zief/Werner/Bradistilov (1998) or Cummins/Doherty/Lo (1999).

[4] For an approach to measure the (re)insurance markets' capacity for covering catastrophe risks see Cummins/Doherty/Lo (1999).

[5] See Swiss Re (1999).

[6] For the structure of recent cat bonds see Doherty (1997a), for the case of non-indexed cat bonds see also Bantwal/Kunreuther (2000).

future financial distress, leads to higher expected costs to complete such projects. The simple strategy of holding "capital inventory" to avoid these costs is usually very costly for emerging economies, making contingent capital an interesting tool especially for emerging countries.

In this paper we will analyze the usefulness of indexed cat bonds from an insurance demand theory point of view, more precisely from the perspective of a primary insurance firm, that has to choose between traditional reinsurance and the issuance of a cat bond. It must be mentioned, however, that of course this kind of analysis in principle also applies to a sovereign government's risk management decision, as long as both options are available.

As was mentioned above, the existence of ART-products is normally explained by its ability to (partly) close the capacity gap of the insurance supply, especially in terms of reinsurance.[7] This line of reasoning is, however, not entirely convincing, since additional capacities could also be acquired by means of extending the level of insurers' (primaries' and reinsurers') equity capital.

In order to explain the increasing relevance of ART, we therefore have to consider the special features of these products.[8] In comparison to traditional reinsurance cover, ART-instruments have to include elements which provide specific advantages for covering certain risks. They have to be analyzed in detail to enhance the understanding of ART's importance and possible usefulness. Surprisingly, insurance economists so far have shown only limited attempts to do this kind of research. Our paper therefore is an endeavor to shed light on the specific advantages and disadvantages of cat bonds as one possible design of ART.

Many papers regarding ART are primarily concerned with the classification and detailed explanation of different possibilities for composing and designing alternative risk transfers. Most of the literature concerned with cat bonds and reinsurance as substitutional risk management instruments is mainly descriptive. Economic analyses in this area have been of a more qualitative kind. The following differences between cat bonds and (re)insurance coverage were identified:[9]

Compared to traditional reinsurance, indexed cat bonds exhibit highly imperfect risk allocation, since they are based on stochastic variables, which are not identical with the losses to be covered. To be of any use

---

7 See e.g. Swiss Re (1996), Kielholz/Durrer (1997), or Müller (1997).

8 See also Jaffee/Russel (1997), who argue that the insurance industry's problems in covering catastrophe risks are caused by the institutional framework, since it limits the incentives for holding sufficiently large amounts of liquid capital, which would be needed to spread such risks over time.

9 See Doherty (1997a), Doherty (1997b), Froot (1997).

they have to be correlated with those losses, but usually cannot be a perfect hedge. Thus a buyer of index-linked coverage always has to face the so-called basis risk.

On the other hand, while an insured usually is more or less in a position to influence the loss distribution, index-linked coverage can be based on an underlying stochastic which cannot be controlled or heavily influenced by the buyer. Thus indexed cat bonds provide a solution for the problem of moral hazard, which can be solved in insurance contracts only by incorporating monitoring or coinsurance provisions.

An additional advantage of cat bonds can be seen in the presumably low transaction costs related to this kind of coverage in comparison to insurance or reinsurance products. Insurance coverage usually incurs considerable costs of acquisition, monitoring and loss adjustment, all of which can be reduced or spared by making use of the financial markets. Furthermore, cat bonds are only weakly correlated with market risk, implying that in perfect financial markets these securities could be sold at a price including just small risk premiums.[10]

To our knowledge there has been no literature so far which – in light of the above-mentioned advantages and disadvantages of both instruments – deals with modelling the simultaneous demand for (re)insurance and index-linked catastrophe risk coverage. With this paper therefore we try to take a first step in this direction by tackling the trade-off between basis risk on the one hand and higher transaction cost on the other.[11] For this purpose we consider the case of a primary insurer facing a catastrophic risk that endangers its insured portfolio. To cover the risk there are two possible opportunities: The primary can buy traditional reinsurance as well as coverage provided by the issuance of an indexed cat bond. The index which serves as a trigger mechanism could be a measure for the extent of a natural disaster in a certain area, for instance the intensity of an earthquake measured according to the Richter scale, or the insurance industry's cumulative losses for that area.

Since coverage based on this kind of index can only serve as an imperfect hedge, it raises the problem of basis risk for the primary insurer. On the other hand it is cheaper, a fact that we incorporate in our model by

[10] See e.g. Litzenberger/Beaglehole/Reynolds (1996) or Lewis/Davis (1998). The profitability of ART-products traded on financial markets so far significantly exceeded the risk free interest rate. This is usually explained by pointing out that high returns were necessary to attract investors to this kind of transactions.

[11] Moral hazard in the reinsurance relationship thus is not considered in this paper.

assuming that the index-linked coverage is sold at an actuarially fair rate,[12] while reinsurance premiums exceed expected losses.

Using the described framework the paper is organized as follows: In section 2 we first introduce our model and then consider the case of a primary who has to choose exclusively the optimal index-linked coverage. We derive the very plausible result that the primary will buy the more coverage the better the cat bond is linked to its insured portfolio, i.e. the less basis risk it would have to face. The more interesting case of ART as well as reinsurance being available to cover the catastrophe risk will be considered in section 3. It can be shown that the availability of cat bonds changes the structure of the optimal indemnity function in an interesting way: it leads to an increase of indemnity for small losses and a decrease of indemnity for large losses. In section 4 we summarize and discuss our results and give an outlook on possibilities for future research.

2. THE MODEL

To become familiar with the model employed in this paper we first analyze the demand for index-linked coverage for the case that reinsurance is not available. We especially want to investigate the impact of a change in basis risk.

A risk-averse[13] primary insurer faces stochastic losses X from an insured portfolio. It considers buying index-linked coverage A, which would be triggered with probability $\overline{p}$. Since this kind of product is usually defined discretely, we can – without major loss of generality – concentrate on the simple case of a stochastic variable with only the two possible outcomes 0 and A. The primary receives the payment, if an exogenous trigger variable Y which is correlated with X reaches a certain level $\overline{y}$.

The correlation between X and Y is expressed by means of the conditional probabilities

$$p(x) := P\{Y \geq \overline{y} | X = x\}$$

[12] The assumption that index-linked coverage is supplied at an actuarial fair rate, is not necessary for our results, but it simplifies the argumentation. Crucial in this context is only that for a given price of reinsurance the alternative coverage is not prohibitively expensive.

[13] For a motivation of risk-averse behavior in entrepreneurial decisions see e.g. Nell/Richter (1996).

(if a certain outcome is not specified we also write $p(X)$).

With these definitions obviously: $\overline{p} = E[p(X)]$.[14]

Consider, for a moment, the following problem: Without any further restrictions, construct an index-linked product with two possible outcomes that is optimal in terms of risk allocation. This product would have to be designed in such a way that the payment A is triggered with probability $p(x) = 0$ for losses up to a certain level, but that it is triggered with certainty if X reaches or exceeds this level. This is due to the feature of decreasing marginal utility, which characterizes a risk-averse decision-maker's von Neumann Morgenstern utility function.

A situation like this, however, is conceivable only if the coverage can be tied directly to X. But then the product would suffer exactly the same moral hazard problems as traditional reinsurance. Since we want to concentrate on instruments that eliminate especially these problems by connecting the coverage to an exogenous index, the situation mentioned above can just be seen as a limiting case for our analysis.

The other extreme case is an index-linked coverage which turns out to be completely useless in terms of risk allocation: If the conditional trigger probability $p(x)$ does not depend on x ($p(x) \equiv \overline{p}$), the primary cannot reduce the risk from its portfolio by issuing a cat bond. So it would simply worsen its situation by buying additional risk.

Naturally, general statements about the function $p(x)$ cannot be made. To keep our argument as general as possible, we only assume that $p(x)$ vanishes for sufficiently small x, and that on the other hand $p(x) = 1$ for sufficiently large losses, and finally that there is an area where the trigger probability is strictly between 0 and 1 and increasing. To formalize this, we say that potential levels of loss $x_1 < x_2$ exist such that

$$\begin{array}{ll} p(x) = 0 & x \leq x_1 \\ 0 < p(x) < 1 & x_1 < x < x_2 \, . \\ p(x) = 1 & x \geq x_2 \end{array} \qquad (1)$$

$p(x)$ is assumed to be differentiable with:

$$p'(x) > 0 \quad \text{for } x_1 < x < x_2 \, . \qquad (2)$$

Fig. 1 shows an exemplary shape of $p(x)$.

---

[14] In the following $E[\cdot]$ always denotes the expectation with regard to the distribution of X.

Figure 1

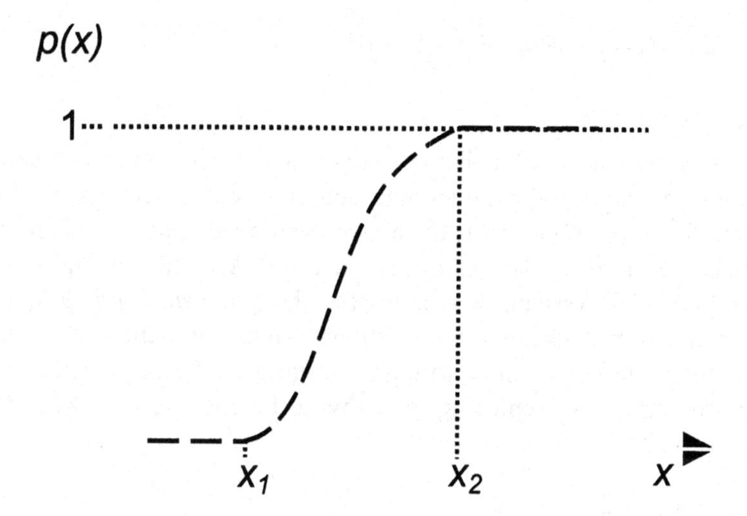

We assume that the index-linked coverage does not cause any transaction cost, and that it is sold on a competitive market at a rate that equals the expected payment. u_1 denotes the (three times continuously differentiable) primary's utility function. It is characterized by $u_1' > 0$ and $u_1'' < 0$, since the primary is risk-averse.

The optimal index-linked coverage in this framework is a solution to the following optimization problem:

$$\max_A E[p(X) \cdot u_1(W_1 - X - \overline{p} \cdot A + A) + (1 - p(X)) \cdot u_1(W_1 - X - \overline{p} \cdot A)].$$

As a first order condition for an interior solution we get

$$\begin{aligned} &\overline{p} \cdot E[(1 - p(X)) \cdot u_1'(W_1 - X - \overline{p} \cdot A)] \\ &\quad = (1 - \overline{p}) \cdot E[p(X) \cdot u_1'(W_1 - X - \overline{p} \cdot A + A)]. \end{aligned} \tag{3}$$

Considering the above-mentioned case that in contrast to our assumptions *p(x)* is independent of *x*, we see that such a cat bond cannot be attractive because in this situation condition (3) would be

$$\overline{p}\cdot(1-\overline{p})\cdot E[u_1'(W_1-X-\overline{p}\cdot A)]$$
$$=\overline{p}\cdot(1-\overline{p})\cdot E[u_1'(W_1-X-\overline{p}\cdot A+A)],$$

implying $A=0$.

To analyze the impact of a change, namely a reduction, in basis risk on the optimal coverage, we examine the consequences of ceteris paribus varying the function *p(x)* towards the above-mentioned situation where the index-linked coverage can be tied directly to *X*. We keep the unconditional trigger probability $\overline{p}$ constant, which justifies the term *ceteris paribus*, and consider a transformation of the conditional trigger probability function that shifts the probability weight to higher values of *x*. More precisely, we consider the effect of replacing *p(x)* by a function $\tilde{p}(x)$ with the properties (1), (2),

$$E[\tilde{p}(X)]=E[p(X)]=\overline{p},$$

and

$$\tilde{p}(x)\leq p(x)\ \ \forall x\leq x_3 \ \text{ und } \ \tilde{p}(x)\geq p(x)\ \forall x\geq x_3 .$$

for an $x_3\in(x_1,x_2)$. To exclude trivial cases we assume $P\{X=x:\tilde{p}(x)\neq p(x)\}>0$.

The idea behind this is that in our setting a product with the same unconditional trigger probability, that is less likely to be triggered for low levels of actual losses but more likely to be triggered for higher losses, means a better fit to the primary's portfolio.

(3) can be reformulated as

$$\overline{p}\cdot E[u_1'(W_1-X-\overline{p}\cdot A)]$$
$$=E[p(X)\cdot\{(1-\overline{p})\cdot u_1'(W_1-X-\overline{p}\cdot A+A)\ +\overline{p}\cdot u_1'(W_1-X-\overline{p}\cdot A)\}].$$

If – starting from an optimal solution – the trigger probability function is transformed in the way described above, the marginal utility levels for large amounts of losses are weighed more heavily. Since u_1' is strictly decreasing, we get:

$$E[p(X)\cdot\{(1-\overline{p})\cdot u_1'(W_1-X-\overline{p}\cdot A+A)\ +\overline{p}\cdot u_1'(W_1-X-\overline{p}\cdot A)\}]$$
$$< E[\widetilde{p}(X)\cdot\{(1-\overline{p})\cdot u_1'(W_1-X-\overline{p}\cdot A+A)\ +\overline{p}\cdot u_1'(W_1-X-\overline{p}\cdot A)\}]$$

and therefore

$$\overline{p}\cdot E[(1-\widetilde{p}(X))\cdot u_1'(W_1-X-\overline{p}\cdot A)]$$
$$< (1-\overline{p})\cdot E[\widetilde{p}(X)\cdot u_1'(W_1-X-\overline{p}\cdot A+A)].$$

In order to fulfill condition (3) again after the variation of $p(\cdot)$, A has to be increased. This result is obviously plausible: All other things equal, the primary will buy the more index-linked coverage the better it fits for compensating the losses from its original risk, i.e. the better the hedge is.

3. THE OPTIMAL RISK MANAGEMENT MIX

We now turn to the analysis of a simultaneous decision on index-linked coverage and reinsurance. As mentioned, we assume that the issuance of indexed cat bonds is cheaper than traditional reinsurance. According to our model cat bonds are traded in a perfect market, and their price equals the expected pay-out. Reinsurance contracting incurs additional costs, which we restrict to the (implicit) cost arising from the reinsurer's risk aversion. However, results similar to those in this paper can be derived by incorporating other types of costs.

The reinsurance premium is denoted by P_2, *I(x)* denotes the indemnity function and u_2 the concave (and three times continuously differentiable) reinsurer's utility function.

We derive Pareto-optimal solutions according to:

$$\max_{I(\cdot),A}\ \alpha\cdot E[p(X)\cdot u_1(W_1-X-\overline{p}\cdot A-P_2+A+I(X))$$
$$+(1-p(X))\cdot u_1(W_1-X-\overline{p}\cdot A-P_2+I(X))] \qquad (4)$$
$$+\beta\cdot E[u_2(W_2+P_2-I(X))].$$

Using the Euler-Lagrange equation[15] the following first order conditions can be derived[16]

$$\begin{aligned} &\alpha \cdot p(x) \cdot u_1'(W_1 - x - \overline{p} \cdot A - P_2 + A + I(x)) \\ &+ \alpha \cdot (1 - p(x)) \cdot u_1'(W_1 - x - \overline{p} \cdot A - P_2 + I(x)) \\ &= \beta \cdot u_2'(W_2 + P_2 - I(x)) \qquad \forall x \end{aligned} \tag{5}$$

or

$$\frac{p(x) \cdot u_1'(W_1 - x - \overline{p} \cdot A - P_2 + A + I(x))}{u_2'(W_2 + P_2 - I(x))} + \frac{(1 - p(x)) \cdot u_1'(W_1 - x - \overline{p} \cdot A - P_2 + I(x))}{u_2'(W_2 + P_2 - I(x))} = \frac{\beta}{\alpha} \quad \forall x,$$

showing the well-known result, that in a Pareto-optimum the marginal rate of substitution is constant.

Furthermore we derive

$$\begin{aligned} &\alpha \cdot E[p(X) \cdot (1 - \overline{p}) \cdot u_1'(W_1 - X - \overline{p} \cdot A - P_2 + A + I(X)) \\ &\quad - (1 - p(X)) \cdot \overline{p} \cdot u_1'(W_1 - X - \overline{p} \cdot A - P_2 + I(X))] = 0. \end{aligned}$$

As a point of reference we compare the optimization problem without the possibility of index-linked coverage:

$$\begin{aligned} \max_{I(\cdot)} \quad &\alpha \cdot E[u_1(W_1 - X - P_2 \\ &+ I(X))] + \beta \cdot E[u_2(W_2 + P_2 - I(X))] \end{aligned} \tag{6}$$

---

[15] See e.g. Heuser (1995), p. 423.

[16] If the reinsurer is risk-neutral, an optimal solution is given by *I(x)=x* and *A=0*. This solution is fairly plausible since under these circumstances reinsurance would be more attractive than index-linked coverage. It could be sold at the same rate and would (of course) not cause any basis risk.

This means we assume the same reinsurance budget for this case as for the situation with cat bonds, which enables us to concentrate on the implications the availability of index-linked coverage has on the *structure* of an ideal reinsurance contract. We do not analyze the impact on the budget spent on reinsurance.

The optimal indemnity function with regard to (6) is defined by

$$\alpha \cdot u_1'(W_1 - x - P_2 + I(x)) = \beta \cdot u_2'(W_2 + P_2 - I(x)) \quad \forall\, x. \tag{7}$$

In the following $I_0^*(\cdot)$ denotes the solution of (6) and $I_I^*(\cdot)$ is the optimal indemnity function from (4).

For a given level of losses x $I_I^*(x)$ will be smaller than $I_0^*(x)$, if the left hand side of (5) is smaller than the left hand side of (7), both evaluated at $I_0^*(x)$. This condition is obviously fulfilled for sufficiently large values of x, respectively $p(x)$.

It must be mentioned, however, that the optimal indemnity function, if index-linked coverage is available, does not assign less indemnity to every level of the loss. This can be seen for the values of x with $p(x) = 0$, since for these x the left hand side in (5) is

$$\alpha \cdot u_1'(W_1 - x - \overline{p} \cdot A - P_2 + I(x)).$$

This expression (for $A > 0$) exceeds $\alpha \cdot u_1'(W_1 - x - P_2 + I(x))$, such that $I_I^*(x) > I_0^*(x)$.

To measure the extent of individual risk aversion we use the Arrow-Pratt measure of absolute risk aversion:

$$r_1(W) := -\frac{u_1''(W)}{u_1'(W)}.$$

In the case of constant or decreasing absolute risk aversion ($\Rightarrow u_1''' > 0$), we can derive the result $I_I^*(x) > I_0^*(x)$ for x with $p(x) \leq \overline{p}$ from the convexity of u_1':

$$\alpha \cdot p(x) \cdot u_1'(W_1 - x - \overline{p} \cdot A - P_2 + A + I(x))$$
$$+ \alpha \cdot (1 - p(x)) \cdot u_1'(W_1 - x - \overline{p} \cdot A - P_2 + I(x))$$
$$\geq \alpha \cdot \overline{p} \cdot u_1'(W_1 - x - \overline{p} \cdot A - P_2 + A + I(x))$$
$$+ \alpha \cdot (1 - \overline{p}) \cdot u_1'(W_1 - x - \overline{p} \cdot A - P_2 + I(x))$$
$$> \alpha \cdot u_1'(W_1 - x - P_2 + I(x)).$$

That the optimal reinsurance indemnity for small losses is larger in a situation where index-linked coverage is available, compared to the model without cat bonds, can be explained quite easily: for small x the effect prevails, that the cost of the index-linked product increases the marginal utility of the reinsurance coverage.

To find out more about the optimal indemnity function we consider the slope of $I_I^*(x)$. Applying the implicit function theorem to (5) we get (where $W_A := W_1 - x - \overline{p} \cdot A - P_2 + A + I_I^*(x)$, $W_B := W_1 - x - \overline{p} \cdot A - P_2 + I_I^*(x)$, and $W_C := W_2 + P_2 - I_I^*(x)$)

$$\frac{dI_I^*(x)}{dx} = \frac{\alpha \cdot p(x) \cdot u_1''(W_A) + \alpha \cdot (1 - p(x)) \cdot u_1''(W_B)}{\alpha \cdot p(x) \cdot u_1''(W_A) + \alpha \cdot (1 - p(x)) \cdot u_1''(W_B) + \beta \cdot u_2''(W_C)}$$
$$- \frac{\alpha \cdot p'(x) \cdot [u_1'(W_A) - u_1'(W_B)]}{\alpha \cdot p(x) \cdot u_1''(W_A) + \alpha \cdot (1 - p(x)) \cdot u_1''(W_B) + \beta \cdot u_2''(W_C)}. \tag{8}$$

The first expression in (8) is positive and smaller than 1, the second is a deduction which can be positive only if $A > 0$. Note that the optimal indemnity function can be decreasing, especially if the function *p(x)* is very steep.

If index-linked coverage is not available or not attractive, the slope of the optimal indemnity function is:

$$\frac{dI_0^*}{dx} = \frac{\alpha \cdot u_1''(W_1 - x - P_2 + I_0^*(x))}{\alpha \cdot u_1''(W_1 - x - P_2 + I_0^*(x)) + \beta \cdot u_2''(W_2 + P_2 - I_0^*(x))}.$$

Taking into account (7) one derives[17]

$$\frac{dI_0^*}{dx} = \frac{r_1(W_1 - x - P_2 + I_0^*(x))}{r_1(W_1 - x - P_2 + I_0^*(x)) + r_2(W_2 + P_2 - I_0^*(x))},$$

where r_1 and r_2 are the primary's and the reinsurer's absolute risk aversion coefficients. If the Arrow-Pratt measures of both of them are constant ($r_1 \equiv a,\ r_2 \equiv b$), $I_0^*(x)$ is a linear function:

$$\frac{dI_0^*}{dx} = \frac{a}{a+b} \tag{9}$$

Now consider again the indemnity function for the situation with index-linked coverage. Dealing with constant absolute risk aversion, we can, without loss of generality, use the utility functions $u_1(W) = -\frac{1}{a} \cdot e^{-a \cdot W}$ and $u_2(W) = -\frac{1}{b} \cdot e^{-b \cdot W}$.[18] For this specific case (8) is of the form

$$\frac{dI_I^*(x)}{dx} = \frac{-\alpha \cdot a \cdot e^{-a \cdot I_I^*(x)} \cdot e^{-a \cdot (W_1 - x - \bar{p} \cdot A - P_2)} \cdot [p(x) \cdot e^{-a \cdot A} + (1 - p(x))] - \ldots}{-\alpha \cdot a \cdot e^{-a \cdot I_I^*(x)} \cdot e^{-a \cdot (W_1 - x - \bar{p} \cdot A - P_2)} \cdot [p(x) \cdot e^{-a \cdot A} + (1 - p(x))] - \ldots}$$

$$\frac{\ldots \alpha \cdot p'(x) \cdot e^{-a \cdot I_I^*(x)} \cdot e^{-a \cdot (W_1 - x - \bar{p} \cdot A - P_2)} \cdot [e^{-a \cdot A} - 1]}{\ldots \beta \cdot b \cdot e^{b \cdot I_I^*(x)} \cdot e^{-b \cdot (W_2 + P_2)}}. \tag{10}$$

From (5) follows

$$\alpha \cdot e^{-a \cdot I_I^*(x)} \cdot e^{-a \cdot (W_1 - x - \bar{p} \cdot A - P_2)} \cdot [p(x) \cdot e^{-a \cdot A} + (1 - p(x))]$$
$$= \beta \cdot e^{b \cdot I_I^*(x)} \cdot e^{-b(W_2 + P_2)},$$

such that (10) can be simplified to

---

[17] For this well-known result from the theory of optimal risk-sharing see e.g. Arrow (1963). The fundamental work on the features of Pareto-optimal risk-sharing rules goes back to Borch (1960). See also Wilson (1968), Borch (1968), Raviv (1979) or Bühlmann/Jewell (1979).

[18] See e.g. Pratt (1964), or Bamberg/Spremann (1981).

$$\frac{dI_I^*(x)}{dx} = \frac{a}{a+b} - \frac{p'(x)\cdot[1-e^{-a\cdot A}]}{(a+b)\cdot[p(x)\cdot e^{-a\cdot A} + (1-p(x))]} \tag{11}$$

By comparing this result to (9) we see that $I_I^*(x)$ and $I_0^*(x)$ are parallel if $p'(x)$ vanishes. This is the case for $x \in [0, x_1)$ and $x \in [x_2, \infty)$. Elsewhere $I_I^*(x)$ is less steep than $I_0^*(x)$.

Figure 2

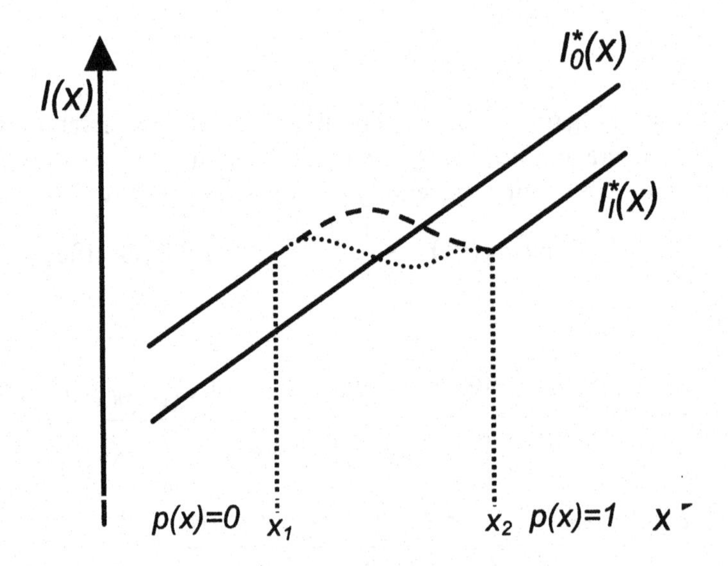

Using (5) and (7) we can also derive an explicit formulation for the connection between the optimal indemnity functions both with and without the supply of index linked coverage:

$$I_I^*(x) = I_0^*(x) + \frac{a}{a+b}\cdot \overline{p}\cdot A + \frac{\ln[p(x)\cdot e^{-a\cdot A} + (1-p(x))]}{a+b}$$

The difference between $I_I^*(x)$ and $I_0^*(x)$ is

$$\left|I_I^*(x) - I_0^*(x)\right| = \frac{a}{a+b}\cdot \overline{p}\cdot A$$

where $p(x)=0$, and

$$\left|I_I^*(x)-I_0^*(x)\right|=\frac{a}{a+b}\cdot(1-\overline{p})\cdot A$$

for $p(x)=1$.

Our central result is that the existence of catastrophe index-linked securities affects the structure of the reinsurance demand: the coverage and thus the indemnity payments increase for smaller losses and decrease for larger losses. The optimal mix of risk allocation instruments thus entails that smaller losses are mainly covered by reinsurance contracts, while larger losses are rather covered by index-linked securities. This confirms the assessment often stated by insurance practitioners, that cat bonds are mainly useful for covering extremely large losses.

This result was derived by assuming the same reinsurance budget in the case with as well as in the case without indexed cat bonds. So we compared the optimal structure of the reinsurance contracts in both situations, but we did not make a statement about the total effect the availability of index-linked coverage has on the amount of reinsurance indemnity for different levels of losses. Of course one would expect a decrease in the demand of reinsurance measured in terms of the reinsurance premium when indexed cat bonds are introduced. It should be kept in mind, though, that one conclusion from our analysis is the following: depending on whether the latter effect predominates, there might be areas where the optimal reinsurance coverage is raised after index-linked coverage comes into play.

4. CONCLUSION

An insurer has at least two important alternatives for covering catastrophic risks: contracting reinsurance coverage or buying index-linked coverage. We analyzed the optimal mix of these two instruments. We have shown that there are strong interdependencies, because both means influence each other heavily with respect to their efficiency.

Obviously, the demand for cat bonds can only be explained via imperfections in the reinsurance market, since cat bonds always result in a basis risk for the insurer. The demand for index-linked coverage cannot be advantageous if reinsurance coverage is offered at fair prices. This implies that transaction costs and/or moral hazard problems in the reinsurance contracts are a conditio sine qua non for the attractiveness of cat bonds. It follows that the demand decision has to balance the trade-off between the basis risk, which is an unavoidable element of indexed cat bonds, on the one hand, and transactions costs and/or moral hazard, which stem from reinsurance, on the other hand.

In our paper we ignored moral hazard and concentrated on the transaction cost problems: Thus we assumed that cat bonds are offered at fair rates, while reinsurance is sold at premiums which exceed expected losses.

We show that, given a certain reinsurance budget, the existence of catastrophe index-linked securities changes the structure of the demand for reinsurance: the coverage and thus the indemnity payments increase for smaller losses and decrease for larger losses. Under realistic assumptions about the risk aversion of the insurer (decreasing or constant absolute risk aversion), the optimal mix of risk allocation instruments thus entails that smaller losses are mainly covered by reinsurance contracts, while larger losses are rather covered by catastrophe index-linked securities. The explanation for this result is that cat bonds imply an additional stochastic element. The parameters of the optimal reinsurance contract therefore change: the coverage for small losses, which imply a small probability that the cat bond is triggered, increases, while the coverage for large losses, which imply a large probability that the cat bond is triggered, is reduced.

To the best of our knowledge our paper is the first which rigorously analyzes the interdependencies between cat bonds and traditional reinsurance. It is a first attempt to tackle the problems in this field. Several important questions must be left for future research. For example, we have dealt with transaction costs only implicitly by assuming risk aversion of the reinsurer. Loss-dependent transaction costs would be an important modification of our model, since one could derive results concerning the connection between the optimal deductibles and the use of cat bonds. Furthermore, a very important step should include moral hazard problems into the analysis.

REFERENCES

Arrow, K.J. 1963. "Uncertainty and the Welfare Economics of Medical Care." *American Economic Review* 53: 941-973.

Bamberg, G., Spremann, K. 1981. "Implications of Constant Risk Aversion." *Zeitschrift für Operations Research* 25: 205-224.

Bantwal, V.J., and H.C. Kunreuther. 2000. "A Cat Bond Premium Puzzle?." *Journal of Psychology and Financial Markets* 1: 76-91.

Borch, K. 1960. "The Safety Loading of Reinsurance Premiums." *Skandinavisk Aktuarietidskrift* 43: 163-184.

Borch, K. 1968. "General Equilibrium in the Economics of Uncertainty." In *Risk and Uncertainty*, edited by K. Borch, and J. Mossin. London: 247-258.

Bühlmann, H., and W.S. Jewell. 1979. "Optimal Risk Exchanges." *Astin Bulletin* 10: 243-262.

Cholnoky, T.V., J.H. Zief, E.A. Werner, and R.S. Bradistilov. 1998. "Securitization of Insurance Risk – A New Frontier." Goldman Sachs Investment Research.

Croson, D., and A. Richter. 1999. "Sovereign Cat Bonds and Infrastructure Project Financing." Working Paper 99-05-25. Wharton Risk Management and Decision Processes Center, University of Pennsylvania.

Cummins, J.D., N.A. Doherty, and A. Lo. 1999. "Can insurers pay for the 'big one'? Measuring the capacity of an insurance market to respond to catastrophic losses." Working Paper. The Wharton School, University of Pennsylvania, Philadelphia, PA.

Cummins, J.D., C.M. Lewis, and R.D. Phillips. 1999. "Pricing Excess-of-Loss Reinsurance Contracts against Catastrophic Loss." In *The Financing of Catastrophe Risk*, edited by K.A. Froot, Chicago: 93-147.

Doherty, N.A. 1997a. "Financial Innovation for Financing and Hedging Catastrophe Risk." In *Financial Risk Management for Natural Catastrophes*, edited by N.R. Britton, and J. Oliver, Brisbane: 191-209.

Doherty, N.A. 1997b. "Innovations in Managing Catastrophe Risk." *The Journal of Risk and Insurance* 64: 713-718.

Froot, K.A. 1997. "The Limited Financing of Catastrophe Risk: An Overview." Working Paper 6025, National Bureau of Economic Research (NBER) Working Paper Series, Cambridge.

Heuser, H. 1995. *Lehrbuch der Analysis - Teil 2*, 9th ed., Stuttgart.

Jaffee, D.M., and T. Russell. 1997. "Catastrophe Reinsurance, Capital Markets and Uninsurable Risks." *The Journal of Risk and Insurance* 64: 205-230.

Kielholz, W., and A. Durrer. 1997: "Insurance Derivatives and Securitization: New Hedging Perspectives for the US Cat Insurance Market." *The Geneva Papers on Risk and Insurance* 22: 3-16.

Lewis, C.M., and P.O. Davis. 1998. "Capital Market Instruments for Financing Catastrophe Risk: New Directions?" Working Paper.

Litzenberger, R.H., D.R. Beaglehole, and C.E. Reynolds. 1996. "Assessing Catastrophe Reinsurance-Linked Securities as a New Asset Class." *The Journal of Portfolio Management*, Special Issue: 76-86.

Müller, E. 1997. "Securitisation – Quo Vadis?" *Zeitschrift für Versicherungswesen*: 597-604.

Nell, M., and A. Richter. 1996. "Optimal Liability: The Effects of Risk Aversion, Loaded Insurance Premiums, and the Number of Victims." *The Geneva Papers on Risk and Insurance* 21: 240-257.

Pratt, J. W. 1964. "Risk Aversion in the Small and in the Large." *Econometrica* 32: 122-136.

Raviv, A. 1979. "The Design of an Optimal Insurance Policy." *American Economic Review* 69: 84-96.

Swiss Re (Ed). 1996. Insurance derivatives and securitization: New hedging perspectives for the US catastrophe insurance catastrophe market, sigma 5/1996.

Swiss Re (Ed). 1999. Alternative risk transfer (ART) for corporations: a passing fashion or risk management for the 21st century, sigma 2/1999.

Swiss Re (Ed). 2000. Natural catastrophes and man-made disasters in 1999: Storms and earthquakes lead to the second-highest losses in insurance history, sigma 2/2000.

Wilson, R. 1968. "The Theory of Syndicates." *Econometrica* 36, 119-132.

13

Considerations Behind the Compulsory Insurance with Earthquake Coverage

M. Hasan Boduroğlu
Istanbul Technical University
Maslak 80626, Istanbul Turkey

INTRODUCTION

Earthquake insurance protects the insured in the private, commercial and public sector against the losses from the earthquake damage, including property damage and for certain cases bodily injury liability. However, great uncertainties in the assessment of hazard and risks makes insurers reluctant in covering earthquakes and potential insurers in purchasing earthquake insurance.

In the last decades, catastropic earthquakes in the world have put great burden on the insurance and re-insurance industry. The main reason is the rapidly increasing vulnerability by the rapidly increasing population, industrialisation and urbanisation.

Because of long recurrence time for strong earthquakes, prediction efforts in terms of single events as well as probabilistic approaches are poor basis for earthquake insurance.

New approaches to the understanding of earthquake activity provide the tools for delinieating locations of potential earthquake hazard. Recognition of the seismotechtonic pattern in space and time poses constraints on future earthquakes in terms of fault mechanisms and magnitudes. Taking into account source, path, and site effects, predictive modelling of the site dependent strong ground motion provides the basis for quantitative analysis of risk and its evaluation with respect to vulnerability and loss estimation. On these grounds, the calculability of risk will be improved and such the estimation of probable maximum losses and liability solidified.

P.R. Kleindorfer and M.R. Sertel (eds.), Mitigation and Financing of Seismic Risks, 255–260.

1. CURRENT INSURANCE POLICIES WITH EARTHQUAKE COVER IN TURKEY

In Turkey, standard home owners' policies do not include coverage for earthquake. One can get insurance for his home and his personal belongings against fire, explosion, and lightening. Once you get this policy, you could get an additional coverage for earthquakes. There is a standard deduction of 5%. Eighty percent of the damage due to an earthquake is covered by the insurer and the remaining 20% by the home owner. Therefore, when there is damage due to an earthquake, 80% with a deduction of 5% will be paid by the insurer. If the total damage is less than the cover, then 20% is paid by the insured and the remaining part will be paid by the insurer after 5% deduction. The insurance is valid only when there is no earthquake damage neither in the building nor in the neighbouring buildings.

During the last decade Turkey experienced five major earthquakes causing life and property losses. In 1992 Erzincan Earthquake with a magnitude of 6.8 caused 750 million US$ material damage and approximately 653 casualties. Only 1.8% of the total cost had been insured against earthquake. 1995 Dinar Earthquake with a magnitude of 5.9 caused 205 million US$ material damage and 94 casualties, and again a negligible percentage of the total cost had had earthquake coverage. Similar situation is also true for 1998 Adana-Ceyhan Earthquake with a magnitude of 5.9. August 17, 1999 Kocaeli Earthquake with magnitude of 7.4 caused about 4 billion US$ material damage and only about 750 million US$ had been provided by the insurance companies. We should note that this coverage is mainly for the industrial facilities.

Although there is no exact figures are available, only 7% of all insurable property is protected against earthquake in Istanbul. This is a rather low percentage when the vulnerability of the buildings is considered.

2. EXAMPLES IN THE STATE OF CALIFORNIA AND NEW ZEALAND

After the devastating Northridge earthquake of 1994, the Legislature created the California Earthquake Authority (CEA) in 1996 to address critical problems impacting the consumers and insurers. If the insurance company is a member of CEA, then the earthquake insurance offered by insurer is a CEA policy. CEA is a privately funded, publicly managed organisation that provides Californians the ability to protect themselves, their homes and their loved ones from earthquake loss. The CEA is committed to:

- Provide actuarially sound earthquake coverage and make its policies universally available and competitively priced.
- Establish the authority's readiness and capability to handle claims promptly, fairly, and consistently.
- Inform Californians about earthquake risk and their options to reduce that risk.
- Provide tools and incentives for earthquake loss reduction and retrofitting.
- Use the best science available in setting its rates.
- Collaborate with other organisations that operate in the public interest to help achieving the CEA's goals.

EQC, the Earthquake Commission, is New Zealand's primary provider of natural disaster insurance to residential property owners. It insures against earthquake, natural landslip, volcanic eruption, hydrothermal activity, tsunami; in the case of residential land, a storm or flood; or fire caused by any of these. The commission is a Crown entity, wholly, owned by the Government of New Zealand. It has a Natural Disaster Fund for the payment of claims. This is currently worth over NZ$4.3 billion. EQC has also arranged its own overseas reinsurance cover. In case of a catastrophic earthquake and, if that is not enough, the Government is required by law to make up any shortfall.

Home owners who insure their property against fire, pay a premium via their insurance company 5 cent per $100 cover. This premium and investment income, makes up the Natural Disaster Fund. As of 30th of June 1999 the fund was NZ$3.37bn. The fund is predominantly invested in New Zealand Government securities.

EQC is a separate entity from private sector insurance and as such has its own excess levels and other conditions which are detailed in the Earthquake Commission Act 1993.

3. NEW DECREE LAW ABOUT COMPULSORY EARTHQUAKE INSURANCE IN TURKEY

After the devastating Earthquakes of August 17, 1999 and November 12, 1999, there has been some action toward earthquake preparedness. One of them is the decree law issued on December 27,1999 about the

Compulsory Insurance. The law covers all independent dwellings, all the offices in these buildings and all the dwellings constructed by the government or dwellings constructed on credit for the replacement of houses damaged due to disasters. Governmental buildings and buildings in villages are excluded. The Natural Disaster Insurance Council has been established. There is a Board of Governors for the Council. There will be a Natural Catastrophe Insurance Pool and The Board of Governors will try to provide the Pool with reinsurance. The law will be in effect in September 2000 for the buildings having construction permits and constructed after the issuance of this law. After November 2000, there will be no free of charge compensation provided under the Law Number 7269 for the replacement of damaged and destroyed residential properties due to natural disasters.

There are certain difficulties involved with the decree law. One of them is the insurance of the existing dwellings. Are they going to be covered by the compulsory earthquake insurance? What will be the cover and the premium? The decree law states that if the fund is not enough to compensate all the damages involved, then the compensation will be paid proportionately. The government does not guarantee to provide funds in excess of the Natural Catastrophe Insurance Pool resources.

A similar compulsory insurance for motor vehicles has already existed in Turkey. An unofficial study shows that only about 70% of the motor vehicles have had this kind of insurance. The problem is how the government will enforce the compulsory earthquake insurance?

In Turkey, several funds have been established for different purposes such as the fund to support the mass housing projects, and the fund to encourage savings for working people. Some of these funds have been mismanaged and have not served for their purposes. Similar danger exists for the Natural Catastrophe Insurance Fund.

4. COMPULSORY EARTHQUAKE INSURANCE FOR EARTHQUAKE DISASTER MITIGATION

A look at the earthquakes occurred in Turkey during the last decade shows that the reasons for high levels of damage and the loss in human life are inadequate engineering, inadequate design reviews, incorrect construction practices, and inadequate inspection or observation of construction. The list above includes both construction trades and individuals responsible for quality control. To be more precise each reason above has a component related to inadequate knowledge and ability. Other factors which contribute to the reasons above, and therefore to the construction quality and seismic safety can be time and budget constraints on the designer (architect and structural engineer), inspectors, and contractors. For most of the building officials, inspectors, or trades seismic resistance has not been a priority topic, until the big August 17, 1999 Kocaeli Earthquake. In order to improve seismic resistance of buildings, a strict control and inspection of construction is a necessity. A certificate can be issued for building, which has been fully inspected at both levels of design and construction phases. There should be lower premiums for earthquake insurance of buildings with such certificates. Turkish Earthquake Foundation has been issuing certificates for fully inspected buildings such as Atasehir Mass Housing Project of Emlak Bank of Turkey. Discount provided on the premium against earthquake insurance will be a very successful closed loop control of building construction and also a very efficient tool for earthquake damage mitigation.

5. CONCLUSION

In spite of the questions raised above, earthquake insurance should be compulsory by jurisdiction everywhere in Turkey. With compulsory insurance, losses will be reduced by encouraging protective measures through the proper tariffs such as allowances, lower deductibles and higher coverage. Adaptation of government guarantee to make up any shortfall in the Natural Catastrophe Insurance Pool should be considered to make the compulsory insurance to become more attractive for the insurer and the insured.

If a careful control is applied from the initial steps of design to the end of construction process of a building, the underlying reasons for high damage can be eliminated. Therefore, a carefully controlled and inspected building can be expected to survive a strong earthquake with minor or no damage. When such a building is going to be insured, the premiums should be much lower than a building with high vulnerability. A certificate of inspection should be given for these buildings both for the design and construction phases. Depending on this certificate reduction of premiums can be applied.

Who should issue this kind of certificate? In Turkey, Loyd's Foundation gives similar certificates for sea vessels. A single non-governmental authority such as the Turkish Earthquake Foundation can issue such certificates for fully inspected buildings, which would be valid for discounts in premiums for insurance against earthquakes. This type of certification will be the best kind of construction inspection.

REFERENCES

Boduroglu, Hasan. 1999. 'Earthquake Disasters and Big Cities,' Urban Settlements and Natural disasters, Proceedings of UIA Region II Workshop, pp.10-14.
1998. Earthquake Prognostics World Forum on Seismic Safety of Big Cities, Abstracts, organised by Turkish Earthquake Foundation and International Commission on Earthquake Prognostics, Istanbul, Turkey.

14

Derivative Effects of the 1999 Earthquake in Taiwan to U.S. Personal Computer Manufacturers

Ioannis S. Papadakis and William T. Ziemba
University of British Columbia

The September 21, 1999 earthquake in Chichi, Taiwan, rating 7.6 in the Richter scale, had devastating consequences. It left approximately 2,300 people dead, more than 10,000 injured, over 100,000 homeless, and about 120,000 unemployed. The government of Taiwan in its late October 1999 estimates expected the economic cost of the earthquake to be $9.2 billion, including the cost of reconstructing the close to 51,000 buildings destroyed and $1.2 billion in industrial production losses (Baum, 1999b).

Not far from the earthquake epicenter, 68 miles north, lies Hsinchu Science-Based Industrial Park. This industrial park has strategic importance for Taiwanese industry and the international computer / electronics market. Large volumes of computer components are produced in Taiwan: motherboards (more than two thirds of world consumption), notebook displays, and various semiconductor products. In particular, 10% of world's memory chips are produced jointly by the 28 fabrication facilities (fabs) situated in the industrial park. It came to no surprise to computer industry analysts, that within 24 hours from the earthquake the government owned electricity company, Taipower, rationed power to 21 million residential and business customers outside Hsinchu Park in an effort to provide reliable power to fabs.

This study provides an overview of the impact of the Chichi earthquake to Taiwanese fabs. It is stressed that the damages to the Taiwanese semiconductor industry could have been much graver had the earthquake epicenter been closer to Hsinchu, or fabs been less prepared for earthquakes, or Taiwan shown a less decisive and prompt response.

The economic impact of the Chichi earthquake, however, was not confined to Taiwan's territory. The computer industry is globalized and tightly interconnected. Disruptions in semiconductor production have immediate consequences to computer / electronics manufacturers who use

P.R. Kleindorfer and M.R. Sertel (eds.), Mitigation and Financing of Seismic Risks, 261–276.

semiconductors as inputs to their products. This paper pays particular attention to the derivative effects of the Taiwan earthquake to the global Personal Computer (PC) market. In order to assess derivative effects of earthquakes, one needs to have an overview of both the direct disaster impact and the mechanism of economic integration that binds the catastrophe struck region with global markets.

We provide a brief overview of the global PC supply chain system. After our analysis it becomes clear that some PC producers are likely to have been more impacted than others by the Chichi earthquake. In light of the possibility for far more devastating disruptions to semiconductor production by earthquakes in Taiwan, we observe that PC producers with different supply chain systems have different exposure to earthquake risks. In particular, companies which are best adapted to the many decade long trend in the computer industry of declining semiconductor component prices are the most vulnerable to earthquake risks.

The PC industry being a modern global industry defining the trends in supply chain management, derivative effect analysis of the earthquake in Chichi provides an excellent case study of a global market disruption due to a natural disaster. Furthermore, we focus on the risk management needs of PC manufacturers using direct sales. Selling directly via the internet, without the intervention of wholesalers and retailers, is considered by analysts to be the future way of doing business, not only in the PC market but also in most other business sectors.

1. IMPACT OF THE CHICHI EARTHQUAKE TO SEMICONDUCTOR PRODUCTION

Following the September 21, 1999 earthquake in Chichi, Taiwan the international association of semiconductor equipment and materials producers (SEMI) dispatched a team of experts to assess the damages. We draw facts from the authoritative report of this expert team (Sherin and Bartoletti, 1999), but we complement their assessment with reports of announcements by the major Taiwanese manufacturers, which had stronger incentives to demonstrate the soundness of fab preparation for earthquakes and emergency response. Characteristically, most major vendors and buyers of computer components didn't wait until the SEMI report was released on December 1999 (about two months after the earthquake). Sparing no expense, they sent their own expert teams to Taiwan immediately after the earthquake (Cataldo, 1999), in an effort to get a good picture of the production disruption fast.

The SEMI report focused on the following aspects of economic cost: 1) Structural damage of buildings, 2) Lost Work In Progress (WIP), 3)

Wafer Fabrication Tool Damage, 4) Damage to sensitive equipment, 5) Hazardous material releases, and 6) Architectural damage. Additionally, an important cost factor was the revenue Hsinchu Park fabs lost per day due to disruptions in electric power.

Sherin and Bartoletti (1999) observe that as Hsinchu Park is situated 68 miles North of the earthquake epicenter, the destructive power of the quake was attenuated leading to only one fab suffering structural damages and most buildings getting away with minor architectural damages. The damaged fab, however, was situated in "soft" soil and it was constructed earlier than most buildings in Hsinchu Park. It should be noted that buildings collapsed in Taiwan's capital, Taipei, situated further North from epicenter than Hsinchu. Collapsed buildings in Taipei, though, were attributed to poor construction. Attenuation was also the reason why fabrication equipment stayed in place, despite reported poor or nonexistent anchorage. There was no major release of hazardous material used as production liquids and gases.

The authors of the SEMI report considered two types of vibration sensitive equipment likely to be damaged after an earthquake: quartz tubes and lithographic steppers. Even though 90% of quartzware was reported damaged after the earthquake, torn pieces of equipment were promptly replaced. Lithographic steppers had to be recalibrated after the earthquake. Lithographic steppers are critical path workstations, this means production may not resume while they are down.

Varying reports (Dillich, 1999 and Scanlan, 1999) estimate the volume of wafer in production lost to be between 28,000 and 20,000 wafers for Taiwan Semiconductor Manufacturing Company (TSMC), the largest semiconductor manufacturer in Taiwan. United Microelectronics Corporation (UMC), the second largest semiconductor manufacturer in Taiwan, lost about 10% of their total wafers in production during the earthquake.

The most significant production disruption occurred due to unreliable electric power after the earthquake. The tremor caused major damages throughout the power grid of Northern Taiwan and electricity is the major energy source for wafer fabrication. Therefore, lack of electric power results in production shutdown. Even loss of power reliability has adverse effects to semiconductor manufacturing, causing damages to sensitive and expensive machinery.

Taiwan's National Science Council (NSC) having reviewed the situation at Hsinchu after the earthquake, managed to obtain cabinet approval for a directive to Taipower (Taiwan's power company) to assign first priority in restoring power to the industrial park. Two important reasons appear to have influenced NSCs decision. First, unreliable power was placing at risk billion dollars worth of fabrication machinery in

Hsinchu Park. And second, every day production ceased Taiwanese Semiconductor Manufacturers would lose $60 million in revenues and Taiwan would forgo an equivalent amount in exports (Baum 1999a). On the other hand, this meant that about 21 million customers of Taipower outside Hsinchu would be left without electricity for several hours per day according to a rationing scheme during the post disaster weeks.

Indeed, six days after the earthquake most fabs had most of their production capacity operational. TSMC reported that all its production equipment were operational 7 days after the earthquake, even though its throughput speed was about 85% of normal. UMC returned to pre-earthquake production rate in 10 days. The total cost of the earthquake was estimated to $88.2 million for TSMC and $12.5 million for UMC (Dillich, 1999). The overall cost for all companies in Hsinchu was estimated to be $200 - $300 million. Its prime components were lost revenue, and damaged wafers in process, with damaged assets being a small fraction of the cost (typically about 5% for most fabs).

Both TSMC's and UMC's costs were covered by insurance. It was originally feared, however, that the Chichi earthquake dealt a serious blow to Taiwanese semiconductor manufacturer's brand name. 1999 was a bad year for Taiwan's semiconductor industry. Taiwan went through a high-tension period with China, which made clear the possibility of armed conflict. This clearly was not the most pleasant news to foreign investors and major semiconductor purchasers. Moreover, in July of the same year a major electric power failure caused serious disruptions in fab production that led to worldwide increases in memory chip prices (Crothers and Shankland, 1999). The Chichi earthquake appeared to be the straw to break camel's back.

Soon after the earthquake the spot market for memory chips saw sharp increases. Contract prices, a better indicator of the cost to major PC producers, went up by 25% during October, 1999. This price increase topped a 50% rise in prices during the summer of the same year (Veverka, 1999). As PC markets are very competitive, increases in component prices mean that profits to PC producers shrink. Dell Computer Corporation, a major PC producer, announced on October 18 that its earnings for the third quarter of 1999 will be less than originally expected, because of the effect of the earthquake to memory chip prices. This announcement, in turn, sent the price of Dell's stock down the very next day by about 7%.

On the other hand, Hsinchu park fabs recovered very fast, within two weeks of the earthquake. Taiwanese semiconductor and computer component manufacturers boosted production during the post-event period by working round the clock in a show of commitment to their production schedules (Baum, 1999b). And this despite competitors' price increases and wholesalers hoarding of products. The swift and decisive response by the Taiwanese government and semiconductor industry after the earthquake appears to have had a positive interpretation, in effect lowering Taiwan's country risk from the point of view of the worldwide PC market (Baum, 1999a).

It shouldn't be underemphasized however, that the Chichi earthquake was not the worst case scenario of a natural disaster to strike Hsinchu park. The industrial park lies twenty kilometers from the epicenter of a 1935 earthquake, of magnitude 7.1, that killed 3,276 people. A disaster of this magnitude, so close to Hsinchu, would result in extensive structural damages to the fabs, overturning and displacement of production equipment, hazardous material releases, and extensive damages to semiprocessed wafers. Sherin and Bartoletti (1999) estimate the economic cost in this event to be in the order of billions of US dollars and the disruption to last for months. Other opinions (Cataldo, 1999) put the worst case disruption period to be from six months to a year. Table 1 provides key indicators from the SEMI report assessment of the Chichi earthquake and the plausible worst case scenario for semiconductor production.

Table 1: Summary of selected SEMI report results

	September 21, 1999 Event	***Plausible Worst Case Scenario***
Epicenter Distance from Hsinchu Industrial Park	68 miles	12 miles
Predominant Cause of Losses	Electric Power Outage	Structural and Equipment Damage
Cost of Damages to Semiconductor Producers	\$200 - \$300 million	Estimated in billions of dollars
Length of Production Disruption	2 Weeks	Estimated in Many Months

During a prolonged semiconductor production disruption due to a severe earthquake, causing structural damages to fabs, one needs to consider not only the behavior of companies that own the damaged facilities, but also the reaction of their competitors. Competitors may see this disruption period as a window of opportunity for them to increase prices and therefore profits. Alternatively, competing semiconductor producers may deem as their best choice to expand their existing facilities or to rush into building new fabs in order to gain in market share. This latter strategy appears to be a sensible option when global supply is constrained. It should, however, be emphasized that it takes a long time to acquire new semiconductor production equipment and have a new production line operational. Therefore, this option is unlikely to be exercised unless an earthquake causes a disruption that lasts for several months. Dormant production capacity and older technology facilities, though, may be utilized to satisfy demand even for relatively brief production disruptions.

2. PC SUPPLY CHAIN SYSTEMS AND THEIR SENSITIVITY TO COMPONENT PRICE SHOCKS

PC supply chains are very complex, rapidly evolving and, as they tend to redefine the state of the art in supply chain management, they are under active research by academics. We provide a brief overview of current practices of and prevailing theories for PC supply chain management as a basis for a later discussion on catastrophic risk analysis. We start with some fundamental insights and then proceed with the latest research applied to the PC industry.

Fisher (1997), in his parsimonious theory of fit between a product and its supply chain, distinguishes between two general product classes: a) innovative products with highly uncertain demand, achieving high profit margins if successful; and b) functional products with very predictable demand offered by many other competitors and resulting in low profit margins. Innovative products require quick response supply chains, which offer possibly many choices to customers and react very fast to customer preferences after the latter are revealed. Functional products (commodities) require low cost supply chains, which deliver a forecasted production volume with the least input and distribution cost.

This theory anticipates that products start their life in the market as innovative products and after some period they become functional products. The transition of a product from the one class to the other may happen so fast that producers are confused about which supply chain to utilize. Often, the same physical supply chain is used by both functional and innovative products, but handling priorities and management practices differ between product classes. PC supply chains are prime applications of Fisher's theory. It is now clear that a disruption in the supply chain of innovative products deserves different analysis from one affecting functional products.

Another important distinction is between decisions concerning product line development and decisions concerning the number of products to be produced within a product line. The concept of a product line in the PC industry is hard to define crisply. Product lines are offered in a variety of versions (flavors), but practitioners find it easy to differentiate a version of a product line from a new product line.

With respect to choosing the right unit volume of a product line, PC producers use various approaches lying between two extremes: the push system and the pull system. According to the traditional push philosophy, producers manufacture large product volumes achieving economies of scale in production and input acquisition and then push the run in big lots to wholesalers and major retailers. If the production lots are larger than the quantity absorbed by the market, profit cutting promotions using discount sales are in order.

On the other hand, the pull system advocates that through a seamless coordination of retailers, wholesalers, producers (assemblers), and component manufacturers, no production takes place before upstream orders arrive. A customer requiring a product generates an order to the retailer, a retailer then turns to the wholesaler and so on. In some cases to avoid unnecessary delays and miscoordination costs all members of the supply chain are notified immediately after a customer order arrives. The "pull" approach, also called "just in time", generates minimal inventories, requiring the less working capital and having the least risk of loosing the value of inventories. Demonstrably, there are many risks to inventories in the PC industry. The major ones are technical obsolescence and devaluation of PC components.

Curry and Kenney (1999) provide a lucid overview of the global PC supply chain. It is disaggregated, meaning not controlled by a small number of companies. It is complex. And, since many people know how to produce (assemble) PCs, timing of product introduction and product delivery become key factors in performing over the global competition. Globally successful companies use supply chain systems appropriately adapted to their needs. On the one extreme companies like IBM, Compaq, and Hewlett-Packard use a smart version of the push system, whereby only inventories of generic products are manufactured. Which product flavor will go to the vendor's self, is determined by the retailers and assembled at their site or close-by. This approach, called sometimes "channel assembly" is described in Davis's (1993) description of Hewlett-Packard's supply chain system.

Curry and Kenney's (1999) analysis offers rave reviews for the Direct-Sales supply chain coordination system, exemplified by the consistently high performing Dell, and Gateway companies. Direct-Sales is a pull system, whereby wholesalers and retailers are bypassed and product assembly starts after customer orders arrive[1]. Customers place orders using a company's internet site or calling the company's 800 phone number. Two weeks or so after the order, the requested computer is delivered to the customer and possibly installed to his/her premises.

Various techniques to manage product development exist so that the lead time to switch from one product line generation to the next is minimized (Thomke and Reinertsen, 1998). Successful PC producers have achieved remarkable switching times, measured in weeks, by integrating product development, marketing, and supply chain management. In addition, Direct-Sales companies have achieved equally impressive product delivery lead times. It may take less than a week for Dell to produce a PC after order arrival (Curry and Kenney, 1999). This makes it easier for Dell to cease production of one product line and move to a new generation of products.

It should be emphasized that Dell's product delivery time, met by a few competitors, is truly remarkable. PC supply chains are highly dispersed geographically, spanning many continents. Coordinating suppliers is very complicated, not only due to the varying locations and cultures of suppliers, but also because specifications, interfaces, and dominant technologies become obsolete fast and are unpredictable. In this rapidly changing environment formalizing supplier contacts is prohibitively costly. Typically, supplier and PC producer are bound by volume commitments guaranteed by codependence and disciplined by the existence of possible alternative contract parties in the global market. Predicting and satisfying a highly variable demand for innovative products, adds further to supply chain management complexity.

PC producers thrive in the face of complexity by being adaptable. Adaptable PC producers, sometimes called "agile", switch from one product line to the next and from one preferred supplier to another fast and at low cost. Adaptability should not be misconstrued as lack of strategy. A Direct-Sales PC producer, like Dell, capitalizes on Moore's Law, the consistent tendency of the ratio of price by performance to decline for most PC components. Michael Dell, founder of Dell Computer Corporation, hints on his company's ability to realize profits from declining component prices by getting paid first and paying suppliers later (Magretta, 1998). Usually, a producer pays for raw material and after some time collects from customers. This period starting when production

---

[1] Certain companies have started selling direct via the internet preconfigured PCs. We will not use the term Direct-Sales for these companies, because they do not have a negative cash-to-cash cycle.

inputs are purchased and ending when final products are sold is called cash-to-cash cycle. Dell Corporation's cash-to-cash cycle is negative.

Curry and Kenney (1999) have developed a theory based on which the faster the decline in component prices the more the profits of a Direct-Sales PC producer. These profits should be contrasted to losses suffered by competitors using a "push" supply chain system, as inventories loose value. Not surprisingly, when component prices go up the effect is reversed. Table 2 provides an overview of the two basic supply chain management systems and their sensitivity to component price changes.

Table 2: Attributes of the two fundamental supply chain management approaches

Approach *Examples*	***Compatible Capabilities***	***Cash -to- cash cycle***	***Effect on Earnings when Component Prices Change***	
			Prices go up	*Prices go down*
Pull *Dell* *Gateway* *Micron*	Configure to order Direct-Sales	Negative	Negative	Positive
Push *Compaq* *IBM* *HP*	Scale economies in production	Positive	Not Necessarily Negative	Negative due to Inventory Depreciation

Table 2 results are based on the premise that the general tendency in component prices follows a decreasing pattern (Moore's Law holds). Therefore, component price increases are not likely to last for long. During a period when component prices increase, PC producers following the "pull" system see their profits declining, as they guaranty prices during the cash-to-cash cycle and sales volume is affected negatively. In the same scenario, PC producers following the "push" approach benefit from inventory appreciation, but, of course, their sales volume is also affected negatively.

Curry and Kenney (1999) assume that Moore's Law stays fairly robust. Despite some worrying signs (Mann, 2000) about the future prospects of Moore's Law, it appears that price increases would be rare and short-lived at least during the present decade (Moore, 1997). Supply disruptions, however, should be considered in relation to demand. The timing of the Chichi earthquake was the less favorable to PC producers, as it came during a period when supply was constrained and demand was at a seasonal year-end high. Inopportune timing is clearly an important risk factor to consider when earthquake induced supply disruptions are investigated.

Table 3 summarizes the results of selected PC producers during accounting periods when memory prices were affected by derivative price increases in memory chips. Dell and Gateway have a pull approach in their supply chain system. Compaq and IBM follow a push approach. Results for two quarters (third and fourth of 1999) are depicted for Dell. Dell cited memory supply problems as the reason for less than expected earnings during the third quarter of 1999 ending on October 29. Dell was expecting an earnings over revenue ratio of 8.2% (last year's result) for both quarters we review. During the fourth quarter of 1999 Dell was also hit by a reluctance from commercial customers to buy PCs before the risks of Y2K conversion were clarified. The same Y2K effect was believed to have hurt every PC producer selling to commercial customers. In Table 3 the Commercial division of Compaq and the Personal Systems division of IBM (targeting mainly commercial customers) have negative results.

Importantly, Dell recognized that the main reason for reduced earnings during the last quarter of 1999 was its failure to realize revenues from about $300 million worth of innovative products due to memory supply disruptions. Gateway another Direct-Sales PC producer suggested restricted memory supply was the major cause for less than expected profits. Compaq Consumer division appears to have been unaffected by memory prices increases.

On the other hand the relatively minor loss of 1999 and even the real risk of future major losses, due to earthquakes very close to fabs, is unlikely to cause Direct-Sales PC producers to abandon their strategy. Most likely Direct-Sales companies will be interested in continuing their core operations strategy, but at the same time shielding themselves from catastrophic risks using specific to contingency risk management policies.

Table 3: Results of selected PC producers during accounting periods when the Taiwan Earthquake affected memory prices

Amounts in USD millions	***Revenue***	***Change over 98***	***Earnings***	***Earnings over Revenue***	***Reason cited***
Dell 3Q99	6,784	41%	483	7%	*Memory supply*
Dell 4Q99	6,801	32%	486	7%	*Memory supply, Y2K*
Gateway 4Q99	2,451	6%	126	5%	*Memory supply, Y2K*
Compaq *Commerc.* 4Q99	3,133	- 19%	-79	-2.5%	*Y2K*
Compaq *Consumer* 4Q99	1,966	24%	69	3.5%	
IBM *Personal Systems* 4Q99	4,131	-7%	-246	-6%	*Y2K*

3. EARTHQUAKE RISK AND PC SUPPLY CHAIN MANAGEMENT

After a short period of uncertainty, analysts settled down on a consensus opinion that the earthquake in Chichi caused a temporary disruption on the PC markets. On the other hand, the SEMI report, drafted by earthquake specialists, is clear. The risk of a far more devastating earthquake, causing a several month long disruption in computer component markets is real (Sherin and Bartoletti, 1999). What priority does this risk receive compared to other considerations in the turbulent environment PC producers operate? Is this risk condemning for Direct-Sales PC supply chain management or are there ways to achieve the full benefit of Moore's Law by managing earthquake risks?

Earthquake engineers define as frequent a seismic event with an expected recurrence period of 34 years (see discussion on SEAOC standards by Sherin and Bartoletti, 1999). The PC industry is less than 25 years old, living in a fast paced world, where an event occurring less often than every year is unlikely to be called a frequently recurring event. It appears that earthquakes are to PC producers low probability - high consequence (lp-hc) events.

Dell, a well managed - to many a model - company, seemingly didn't have a plan for the 1999 Taiwan earthquake. Two facts concur to this conclusion. First, Dell reported less than expected profits during the last quarter of 1999, by its own admission due high memory prices. Second, it had an incentive to reveal any existing earthquake risk management policy to calm worried investors during the post event period and thereby protect its stock price from falling.

Dell did use a shield against the derivative effect of the Taiwan earthquake. But, apparently, it was not using its long term contracts with suppliers, as was suggested by some investment analysts (Deckmyn, 1999). Dell used its capability of being "agile" to quickly adopt its product configurations to the high memory price environment at the end of 1999. Shortly after the effects of the disaster were felt, Dell promoted and made available products using less memory and having some compensating features, like larger monitors and better disk drives.

It is difficult to estimate product demand for innovative products in the PC industry, therefore signing long term contracts for all inputs of a PC producer is itself a risky plan. Moreover, long term contracts for PC components would likely negate Dell's comparative advantage as a Direct-Sales company to achieve profits when component prices decrease. The prevailing sentiment among earthquake experts is that earthquakes cannot be predicted accurately with respect to timing, location, and magnitude (Geller et al.,1997). Furthermore, for the purposes of supply chain management an earthquake warning would need to arrive early. This added specification makes precise predictions of earthquake induced production disruptions even less likely to be obtained. Probability distributions for seismic events in a location over the span of decades can, however, be assessed fairly accurately.

Decision scientists would not find lack of preparedness for earthquakes and failure to use this probabilistic information surprising. Kunreuther (1995), in an overview of his research results on the failure by homeowners to invest in risk mitigation and to buy insurance against natural disasters, cites among other possible explanations the following: a) decision makers fail to think according to the norms of probability theory, b) risk is a multidimensional problem, and c) people use simplified decision rules (e.g., ignore risks if probability is under a threshold value). On the other hand, Dell, rich in human and financial resources, cannot be directly compared to possibly strapped for cash homeowners. Kunreuther's research, however, is based both on decision maker behavior under real lp-hc dilemmas and under simulated dilemmas during laboratory experiments. Therefore, it suggests that biases in processing lp-hc information are deeply rooted and good predictions of a decision maker's actual behavior. Do these results suggest that PC producers will go on with business as usual without some protection against earthquake risk? To the contrary, decision science research suggests that decision makers are likely to overestimate the probability of a disaster after one occurs (see discussion on availability bias in Kleindorfer, Kunreuther, and Schoemaker 1993, 94-95).

Doherty (1997) provides an overview of existing financial and insurance products for primary insurers covering natural disaster losses. In particular, he classifies four basic strategies: 1) asset hedge, 2) liability hedge, 3) post loss equity re-capitalization, and 4) leverage management. One may imagine similar products offered to PC producers against natural disaster derivative effects. In the current environment of rapid innovation in financial and insurance products, it may also be possible to develop products tailor-made to the needs of Direct-Sales PC producers. The design of these risk transfer instruments deserves special attention and is beyond the scope of this study. We discuss briefly in the following the estimation of the basic risk parameters, the probability and magnitude of derivative losses.

It is technically feasible to estimate probabilistically with adequate accuracy the magnitude of losses in a specific geographic region for a long horizon (usually measured in decades). Moreover, companies specialize in the provision of this type of probability distributions. Data on expected earthquake frequency will be available to any analyst willing to pay the the cost of information. The magnitude of the price shocks on PC components following an earthquake in a region with high concentration of semiconductor production will be more difficult to estimate. For the purposes of contract design one needs to recognize the most important parameters controlling post-event prices. These are:

1. Loss of semiconductor volumes on every day after an event of interest until fabs return to regular production rates. More simplistically, one may consider two parameters: average post event loss of production capacity and average length of disruption.

2. Total world semiconductor production capacity at time of the event of interest.

3. Demand during the post-event period. Historically, this demand is seasonal and increasing on aggregate from year to year.

Of paramount importance to Direct-Sales companies is the risk of a sustained period of growth in component prices (a gradient effect) during the post-event period. The most likely scenario, however, is that observed after the Taiwan earthquake. Spot prices rose to their highest level very fast - to a large extent due to hoarding - driving contract prices up too. After the crisis subsides, prices are likely to follow a downward path to regular levels. Alternatively, repeated price increase shocks may occur, leading to repeated earnings erosion, the vice versa of what occurred during component price wars. It appears to be in the best interest of profit maximizing PC component producers and wholesalers to quickly assess the extend of post-event damages, promptly raise prices, and lower them only after production capacity goes up. On the other hand markets are known to behave counterintuitively, and general equilibrium models may be the best course of action when accuracy is desired.

If one accepts this fast market correction hypothesis, then the most important risk parameter is price increase magnitude. Direct-Sales companies will suffer large losses during market correction. Due to their negative cash-to-cash cycle, however, they are likely to start realizing profits again when prices begin a downward trajectory. Interestingly, according to this scenario, Direct-Sales companies' going back to profits depends on market behavior only and not on the dynamics of productive capacity recovery.

4. CONCLUSION

The September 21, 1999 earthquake in Chichi, Taiwan resulted in an unexpected price increase of an essential component of PCs, memory chips. This 25% price increase was sizable, but not exorbitant. The reason why it proved very hard to manage was that it came in an environment of already restricted supply and repeatedly increasing prices throughout the last part of 1999. The memory price increase appears not to have had a material effect on the earnings of major PC producers following the traditional "push" supply chain management system. Both major PC producers following the "pull" approach, however, reported less than expected earnings due to high memory component prices.

As Dell and Gateway faired very well compared to their competitors, it appears that their configure-to-order and negative cash-to-cash cycles are essentially sound operation management strategies, contributing to their high growth and profitability. Consequently, these strategies need to be continued but complemented by contingency-specific risk management policies. The latter should offer more options to PC producers exercising them during a crisis. Risk management policies should alleviate the burden of high priced production components during an unwanted contingency, leaving PC producers with sufficient cash reserves to pursue growth opportunities. Bankruptcy risk is less of an issue for companies characterized by high growth in earnings, like Dell and Gateway.

Another important consideration arises from the results of the earthquake engineering analysis compiled on behalf of SEMI (Sherin and

Bartoletti, 1999). A plausible worst case scenario for Semiconductor production is a 7.1 magnitude earthquake striking very close to Hsinchu industrial park. In this scenario many production facilities would have material damages and production will be disrupted for many months. PC producers, when deciding on risk management policies, are likely to be interested in a range of relevant scenaria and not just in a repetition of the Chichi event. Significant insights are to be gained by a detailed earthquake risk analysis of PC supply chains. Its results would be useful to the Semiconductor industry, to PC producers, and possibly to insurance companies and financial institutions interested in offering risk transfer instruments to PC producers.

REFERENCES

Baum, Julian. 1999a. "Shaken not stirred" *Far Eastern Economic Review* (October 7).

Baum, Julian. 1999b. "Back on track" *Far Eastern Economic Review* (November 25).

Cataldo, Anthony. 1999. "Taiwan fabs face earthquake risk" *Electronic Engineering Times* (December 6).

Compaq Corporation. 2000. "Compaq announces fourth quarter, full year 1999 results" *Compaq press release* (January 25).

Crothers, Brooke, and Steven Shankland. 1999. "Taiwan quake could disrupt computer, chip markets" *CNET News.com* (September 21).

Curry, James and Martin Kenney. 1999. "Beating the clock: Corporate responses to rapid change in the PC industry" *California Management review*, 42:1(Fall), 8-36.

Davis, Tom. 1993. "Effective supply chain management" *Sloan Management Review* (Summer) 35-45.

Deckmyn, Dominique. 1999. "Memory price increase hits Dell" *Computerworld* (October 25): 91.

Dell Computer Co. 2000. *Annual Report for FY ending January 28, 2000: Form 10-K.* Securities and Exchange Commission, Washington, DC.

Dillich, Sandra. 1999. "Big losses after the quake" *Computer Dealer News* (October 22). Plesman Communications.

Doherty Neil A. 1997. "Financial innovation in the management of catastrophe risk" *Bank of America: Journal of Applied Corporate Finance*, 10:3(Fall): 84-95.

Fisher, Marshall L. 1997. "What is the right supply chain for your product?" *Harvard Business Review* (March-April): 105-116.

Gateway Inc. 2000. *Annual Report for FY 1999: Form 10-K.* Securities and Exchange Commission. Washington, DC.

Geller, Robert J., David D. Jackson, Yan Y. Kagan, and Mulagria Francesco. 1997. "Earthquakes cannot be predicted" *Science*, 275:5306(March 14): 1616.

IBM Corporation. 2000. "IBM announces 1999 fourth-quarter, full-year results" *IBM press release*, January 19.

Kleindorfer, Paul, Howard Kunreuther, and Paul Schoemaker. 1993. *Decision Sciences: An integrative perspective.* Cambridge University Press.

Kunreuther, Howard. 1995. *Protection against low probability high consequence events.* Wharton Center for Risk Management and Decision Processes: Research Paper.

Magretta, Joan. 1998. "The power of virtual integration: A interview with Dell Computer's Michael Dell" *Harvard Business Review* (March-April): 73-84.

Mann, Charles C. 2000. "The end of Moore's law?" *Technology Review*, 103:3(May-June): 42.

Moore, Gordon E. 1997. "The microprocessor: Engine of the technology revolution" *Communications of the Association of Computing Machinery*, 40:2 (February): 112-114.

Scanlan, Sean. 1999. "When the chips are down" *Asian Business* (November).

Sherin, Brian and Stacy Bartoletti. 1999. *Taiwan's 921 Quake: Effect on the semiconductor industry and recommendations for preparing future earthquakes.* Semiconductor Equipment and Materials International (SEMI), Mountain View, CA.

Thomke, Stefan and Donald Reinertsen. 1998. "Agile product development: Managing development flexibility in uncertain environments" *California Management Review*, 41:1 (Fall): 8-30.

Veverka, Mark. 1999. "A DRAM shame" *Barron's* (October 25): 15.

15

Insurance Markets for Earthquake Risks in Turkey:
An Overview

Barbaros Yalçin
MilliRe Corporation

This paper provides an overview of insurance markets in Turkey, with special reference to those aspects of the market related to seismic risks. These markets and their associated regulatory and financial institutions have been undergoing significant changes in the past year, and so this paper may be best considered a snapshot as of the summer of 2000, with on-going changes in legal, financial and actuarial institutions continuing to reshape these markets.

When we look at a map showing the earthquake zones of the planet, it is readily seen that some countries are faced with more earthquake risk than others. Indeed, as other papers in this volume make clear, Turkey is located in one of the most active seismic areas on the planet. Table 1 lists just the major quakes that have occurred in the past century. We see that during the last 106 years, 65 earthquakes occurred in Turkey having a magnitude of 6 or greater. In these earthquakes, 94.558 people have lost their lives, many people have been injured and many buildings demolished and damaged.

One of the most disastrous earthquakes occurred recently, the earthquake in the Marmara Region of 17th August 1999. This event, which is still fresh in our minds, caused a loss of some 10 billion US$ to our economy, with some estimates putting the loss as high as 20 billion US$, not to mention the huge loss of nearly 20,000 dead.

Although the losses caused by major earthquakes to Turkey's economy are not known exactly, these have certainly been large. Generally quoted figures note that the country's average economic loss due to earthquakes has been around 150-200 million US$ annually. Given these losses, there has been a demand for insurance to provide some coverage for the physical damages caused by earthquakes. This coverage is sometimes automatically

P.R. Kleindorfer and M.R. Sertel (eds.), Mitigation and Financing of Seismic Risks, 277–286.

included in an insurance policy and sometimes added to the policy upon agreement of both parties.

Table 1: Earthquakes of Magnitude M $\geq$ 6 in Turkey 1894-1999

Date	Place	Magn.	Deaths	Date	Place	Magn.	Deaths
10.07.1894	İstanbul	7.0	474	23.07.1949	Karaburun	7.0	1
08.11.1901	Erzurum	6.1	500	17.08.1949	Karlıova	7.0	450
28.04.1903	Malazgirt	6.7	2.626	13.08.1951	Kurşunlu	6.9	52
28.04.1903	Patnos	6.3	3.560	18.03.1953	Gönen	7.4	265
4.12.1905	Malatya	6.8	500	16.07.1955	Söke	7.0	23
19.01.1909	Foça	6.0	8	25.04.1957	Fethiye	7.1	67
09.02.1909	Menderes	6.3	500	26.05.1957	Abant	7.1	52
09.08.1912	Mürefte	7.3	216	23.05.1961	Marmaris	6.5	0
03.10.1914	Burdur	7.1	4.000	18.09.1963	Çınarcık	6.3	1
24.01.1916	Tokat	7.1	500	14.06.1964	Malatya	6.0	8
18.11.1919	Soma	6.9	3.000	06.10.1964	Manyas	7.0	23
13.09.1924	Pasinler	6.9	310	18.08.1966	Varto	6.9	2.394
13.09.1924	Horasan	6.9	50	22.07.1967	Adapazarı	7.2	89
18.03.1926	Finike	6.9	27	26.07.1967	Pülümür	6.2	97
31.03.1928	Torbalı	7.0	50	30.07.1967	Akyazı	6.0	0
18.05.1929	Suşehri	6.1	64	03.09.1968	Bartın	6.5	29
06.05.1930	Hakkari	7.2	2.514	14.01.1969	Fethiye	6.2	0
04.01.1935	Erdek	6.7	5	23.03.1969	Demirci	6.1	0
01.05.1935	Digor	6.2	200	25.03.1969	Demirci	6.0	0
19.04.1938	Kırşehir	6.6	149	28.03.1969	Alaşehir	6.6	41
22.09.1939	Dikili	7.1	60	28.03.1970	Gediz	7.2	1.086
26.12.1939	Erzincan	7.9	32.962	12.05.1971	Burdur	6.2	57
20.02.1940	Develi	6.7	37	22.05.1971	Bingöl	6.7	878
23.05.1941	Muğla	6.0	2	27.03.1975	Gelibolu	6.4	7
15.11.1942	Bigadiç	6.1	7	06.09.1975	Lice	6.9	2.385
20.12.1942	Niksar	7.0	3.000	24.11.1976	Çaldıran	7.2	3.840
20.06.1943	Hendek	6.6	336	30.10.1983	Horasan	6.8	1.155
26.11.1943	Tosya	7.2	2.824	07.12.1988	Akyaka	6.9	653
01.02.1944	Gerede	7.2	3.959	13.03.1992	Erzincan	6.8	94
25.06.1943	Gediz	6.2	21	27.06.1998	Ceyhan	6.3	145
06.10.1944	Edremit	7.0	27	17.08.1999	Kocaeli	7.4	17.127*
20.03.1945	Ceyhan	6.0	10	12.11.1999	Düzce	7.2	845*
21.12.1945	Denizli	6.8	190				

Source: Istanbul Technical University Review, December 1999
Supplemented by * = Prime Ministry Crisis Management Center

The "General Conditions" prevailing in the Turkish insurance market related to earthquake losses include the following:

- Motor Third Party Liability, Various Liability, Plate Glass, Machinery Breakdown, Compulsory Personal Accident for Busses and Health do not cover losses related to earthquakes at all.
- In Fire, Marine Hull, Marine Cargo, Motor Own Damage, Personal Accident, Burglary, Electronic Equipment, and Hailstorm insurances, earthquake coverage is given as an additional cover with the agreement of insurer and the insured. In particular, an additional cover is given in Fire policies against the physical damage caused by the earthquake.
- In E.A.R. (Erection All Risks), C.A.R. (Contractors All Risks), Marine Special, Poultry, Livestock and Life insurances, earthquake coverage is included in the given coverage automatically.

Perhaps because of the salience of the August 1999 earthquake, there was significant growth in demand for earthquake coverage in the last half of 1999. Thus, there were additional Earthquake covers on 625,355 Fire and Engineering policies as of 30.06.1999 and the total sum insured in these policies was US$ 98,817,644,000. By 31.12.1999, the number of Fire and Engineering policies including additional Earthquake cover had risen to 1.064.579, and the total sum insured by these policies had risen to US$ 130.692.095.000.

A similar growth took place in dwelling/fire policies following the August 17th event. Table 2 below shows the changes in the total exposures, policies and premiums for dwelling/fire policies with earthquake cover included.

Claims paid by Turkish insurance companies due to recent earthquakes are shown in Table 3 below. From this we see that the Turkish insurance sector has paid an indemnity of around US$ 646 million from the 1992 Erzincan earthquake up to the end of 1999.

In most developed countries, earthquake risks are not a governmental responsibility for industrial and commercial losses resulting from earthquakes. The reason is that insurance professionals are believed to be better able to estimate the probable risks these enterprises face, and loss reduction and risk bearing are therefore appropriately undertaken by the insurance industry in cooperation with the owners of industrial or commercial properties. This internalises loss prevention and risk bearing among private parties. The main problem regarding recovering of loss due to earthquakes arises with household dwellings. Here there are a variety of opinions about how these losses should be covered.

Table 2: Earthquake Cover Included in Fire Policies

Year	Sum Insured (million TL)	Increase %	No. of Policies	Increase %	Premium (million TL)	Increase %	Average Rate ‰
1987	6.920.012	-	137.058	-	6.228	-	0.9
1988	11.802.387	71	224.806	64	9.681	55	0.82
1989	25.141.144	113	263.926	17	14.037	45	0.55
1990	40.461.265	61	409.56	55	19.89	42	0.49
1991	87.506.994	116	536.28	31	49.203	147	0.56
1992	188.146.055	115	660.125	23	82.714	68	0.44
1993	333.476.707	77	716.524	9	174.125	110	0.52
1994	623.037.830	87	796.042	11	626.451	260	1
1995	1.343.353.911	116	676.514	-15	1.100.147	76	0.82
1996	2.670.489.866	99	674.963	-	2.074.358	89	0.78
1997	6.222.517.509	133	645.468	-4	4.744.134	119	0.76
1998	14.802.867.015	138	655.751	2	11.406.098	140	0.77
1999	29.838.829.219	102	663.929	1	24.012.188	111	0.8
2000	61.485.428.461	106	1.027.998	55	55.099.574	130	0.9

Table 3: Claims Paid by Turkish Insurance Companies for Earthquake Losses

Date	Place	Magnitude	Insured Loss
13.03.1992	Erzincan	6.8	US$ 20.0 million
27.06.1998	Ceyhan	6.3	US$ 5.5 million
17.08.1999	Koçaeli	7.4	US$ 570.0 million
12.11.1999	Düzce	7.2	US$ 50.0 million
			US$ 645.5 million

1. WHO WILL COVER THE LOSSES TO DWELLINGS CAUSED BY EARTHQUAKES?

According to one opinion, considering that large earthquakes are infrequent events, with damages not the result of action or inaction by victims, the Government should cover the losses to dwellings caused by earthquakes. Those who advocate this approach also urge that the Government should exercise control locally for emergency response and in implementing building codes and zoning regulations determining the types of buildings and construction to mitigate earthquake risk.

Opposing this view are those who believe that if the Government covers the losses of earthquake victims, building owners' and local authorities may act negligently in failing to take necessary measures to decrease potential losses to structures in earthquake regions. Moreover, the economic situation will be disrupted after a major event, and the Government may not have the ability to provide sufficient support for stricken regions. In this situation,

The Government will then be forced to incur foreign debt, which is a decidedly unfavourable outcome.

Another opinion regarding coverage to losses of dwellings caused by earthquakes is that the Government absolutely should not help victims after the event. The owners of the dwellings who think that they may face a loss, may then buy an insurance cover from private insurance companies, and those who do not buy insurance will bear the loss themselves. If the Government covers a substantial portion of earthquake losses through post-disaster relief, then there will no incentive to purchase private insurance or, indeed, to undertake structural mitigation measures before the fact.

The major argument against this opinion is that many dwelling owners who do not buy an insurance cover may not have the financial wherewithal to repair or reconstruct their dwellings. Accordingly, they may be in a very precarious situation after an event, and the Government may have little choice but to step in with emergency aid.

A third opinion states that dwelling owners should be compelled to buy earthquake insurance. This obligation may be implemented by Government decree or by banks and mortgage institutions if most buildings are covered by mortgages. In our country, however, buying a house using credit from banks is not widespread, so there isn't much insurance made due to credit on mortgage. However, there is currently a large demand for long-term housing credit, which some bankers initiated during the last days of the year 1999. As the economy gains stability, interest rates will decrease and become stable, and its seems likely that buying a house using long-term housing credits from banks may become more widespread in our country, providing additional leverage for the purchase of earthquake insurance.

In the interim, the most important development in Turkey has been the "Decree Law Regarding Compulsory Earthquake Insurance" issued on the Official Newspaper dated 27th December 1999. This Law requires that buildings used for housing (apartments, condominiums) as well as those parts of buildings used as commercial houses, offices and such, must obtain earthquake insurance as of September 2000. According to the 10th Article of the said Decree Law, the cover limits, General Conditions, Tariff and Instructions, as well as the principles underlying the determination of premiums will be determined by the Ministry. According to the General Conditions, Tariff and Instructions of the Compulsory Earthquake Insurance published in Official Newspaper dated 8th September 2000, Compulsory Earthquake insurance policies are to be issued, and cover provided, by the TCIP (Turkish Catastrophe Insurance Pool). Distribution and sale of these policies is to be accomplished by existing insurance companies and their agencies as authorized by TCIP. Existing insurance companies are only the intermediaries in the process of selling the

Compulsory Earthquake policies. All earthquake risk is the ultimate responsibility of TCIP.

With Compulsory Earthquake insurance, material loss and damage of insured buildings caused directly by earthquakes (including fire, explosion and landslides following earthquakes) are covered up to the insured amount. The insured amount is up to a maximum 20 billion TL (about US $ 20,000) for each dwelling (each apartment or independent housing unit in multi-dwelling buildings is covered up to the same maximum amount). If the cost of the dwelling is more than this amount, it is possible to obtain additional cover through facultative insurance for the excess amount from a private insurer.

A simple Tariff has been designed for this Compulsory Earthquake insurance. In this Tariff; buildings are divided into 3 construction types and the country is divided into 5 zones according to the earthquake risk faced in that location, so that 15 different rates are determined. The building's insured amount is determined by the cost of construction per square meter (determined by the Undersecretariat according to the type of building) multiplied by the building's area in square meters. Moreover, a deductible of 2% of the insured amount will be applied to each claim against the Compulsory Earthquake insurance.

Since earthquakes are a catastrophic peril, insuring these risks is extremely complex. Notwithstanding new catastrophe modelling techniques, there are still significant uncertainties associated with predicting the frequency and severity of these events. Therefore, insurance companies providing earthquake cover need considerable risk capital to provide reliable insurance for such events, and this means a lot of money. For example, for the City of Los Angeles, it has been estimated that a large earthquake could cause more than US$ 50 billion in insured losses. As seen in Table 3, the paid insured loss amount for the 17th August earthquake in our country was around US$ 570 million. However, such large losses can bankrupt an insurer, so insurers have found and developed different ways of protecting themselves.

2. INSURANCE COMPANIES' STRATEGIES TO PROTECT THEMSELVES IN COVERING EARTHQUAKE LOSSES

One way is to spread the risk geographically. For this, insurance companies try to have a balanced portfolio by accepting earthquake insurance not only from the risky regions, but also from safer regions.

Another way is to establish limits on earthquake covers. Insurance companies may give earthquake cover with a certain proportion of 100% of the insured amount. And also, in the event of a loss, there may be "deductibles" applied in a certain proportion fixed beforehand. Such coverage limits and deductibles are common in Turkey.

However, even after applying these direct approaches, insurance companies may still have excess exposures. This leads them to seek additional risk transfers through reinsurance. Such reinsurance may take any of the standard contracts or treaty forms common in reinsurance including proportional and non-proportional sharing and Pool agreements. Recently securitizing earthquake risks and transferring the risk to Stock Markets has also become an option. As these are all recognized and well-known means of pooling and transferring direct insurer risks, I will not focus on these. I wish instead to focus on the issue of narrowing of the traditional reinsurance markets and the resulting cost increases of reinsurance for catastrophic events.

The insurance and reinsurance companies had to pay big amounts of indemnification after the catastrophic losses in the world and especially in the U.S.A. (i.e. after the 1992 Iniki and Andrew hurricanes and 1994 Northridge earthquake). After these losses, several companies exited the market, reinsurance capacity in the world decreased and the cost of reinsurance increased, at least for a while. The consequence was that direct property and casualty insurers abstained from insuring catastrophic risks, as they could not easily find reinsurance cover. Some governments, noting these developments, established different kinds of insurance systems, which would create funds in order to cover catastrophic losses. These systems have been in force in the U.S.A. for many years in different forms. At first they were in a federal condition but afterwards each state established its own insurance system. As examples, The NFIP (The National Flood Insurance Program) protecting all the dwellings against flood losses, the FCIP (Federal Crop Insurance Program) covering agricultural products against catastrophic losses and the CEA (The California Earthquake Authority) established to cover the earthquake losses in California may be given. Thus, even in the U.S.A. insurance sector; with its very strong financial strength and which had a surplus of US$ 310 billion as of 31.12.1997, insurance companies have big difficulties in covering catastrophic losses and therefore the Government took some measures regarding this problem.

Consider, for example, the California situation. The huge loss which the 1994 Northridge Earthquake (Magnitude 6.7) caused was the reason for the establishment of the CEA. The CEA is a combination of public-private sector institute managed by the public but financed from several sources as established by Law in order to assure the building owners, tenants, householders and mobile-householders an insurance coverage against earthquake risk with the cooperation of insurance companies. 15 insurance companies licensed to do business in California are members of CEA and provide one layer of risk capital for policies underwritten by the CEA, with other layers covered through state financing and reinsurance. Other insurance companies that are not members of CEA can sell their own earthquake policies.

3. CATASTROPHIC LOSS PROGRAMMES IN OTHER COUNTRIES

Systems like the CEA can also be seen in other countries facing catastrophic losses. The insurance companies in these countries requested the Government's help due to the catastrophic losses of which the number and the cost have been continuously increasing. As an example, " Insurance Indemnity Consortium" has been in practice since 1954 in Spain. It covers all the structural, contents and vehicular losses caused by catastrophic losses, terrorism, revolts, and malicious acts against the Government. This Consortium, which had been under the Ministry of Finance, had many changes over time and since 1991 has been an autonomous institute separate from the Government.

Another example is the "Natural Catastrophe Insurance System" applied in France since 1982. France has faced many catastrophic losses, and all the natural catastrophes are included in the fire policy and covered by the insurance companies. However, the natural catastrophe's risk being so large, securing a cover from the world reinsurance market is difficult, so this system has been given to the management of Caisse Nationale de Reassurance (C.C.R.), a reinsurance company guaranteed by the Government. As can be remembered, the Martin and Luther storms in 1999 and the heavy rain a few days before the New Year caused a big loss in France, reported as around US$ 8-10 billion.

Another example of the cooperation of the insurance companies and the Government in covering catastrophic losses is the Earthquake Commission (EQC) in New Zealand. This Commission was established within the framework of the Earthquake and War Loss Law put into force in 1944. The aim of this Commission is to provide insurance against earthquake, tsunami, land slide, volcanic eruption and the physical damage due to fire caused by them. Initially, insurance cover was given only to commercial risks and excluded dwellings, but since 1990 it covers only dwellings. The householders buying a fire insurance policy from insurance companies in New Zealand have to buy the earthquake cover of the Earthquake Commission. The Earthquake Commission gives a maximum cover of

100.000 NZD to the dwellings and 20.000 NZD to contents. People who want to buy an additional cover above these limits can apply to private insurance companies. If the Commission's resources and the reinsurance cover are not sufficient to cover losses, then the Government pays the excess.

Other examples of private-public partnerships in risk bearing include those of Norway and Japan. Currently studies are being undertaken in Belgium, Italy, Greece and Australia of similar systems in these countries.

4. CATASTROPHE BONDS AND NEW FINANCIAL INSTRUMENTS

Another major innovation in catastrophe insurance has been the design of new financial instruments to allow investors in capital markets to provide additional risk-bearing capital to cover catastrophe losses (see also the Nell and Richter paper in this volume). Investors in stock markets seeing the difficulties of the insurance sector in finding a reinsurance cover, especially after large losses, have introduced securities that trigger payments to the insurance company when large events occur and otherwise act just like regular bonds when no major events occur. In this sense, these securities are potential substitutes for reinsurance. Although the idea of issuing a security for catastrophic losses in stock markets has been known since 1993, investors have become more interested in this idea in the last as prices of these catastrophe securities have risen relative to alternative investments. For example, during the last years, whereas the American Treasury Bonds yielded 6% interest, Catastrophic Bonds' produced yields of around 10-12%.

Investors put their money in some independent financial facility that promises to pay these investors a given interest rate if no event occurs during a specified time period. Investors lose some portion of their principal and agreed interest to insurers (or other holders of these bonds) if some pre-specified trigger event (e.g., an earthquake of magnitude greater than 7.0 within 100 kM of Istanbul) occurs during a specified time period. The independent facility invests the money in the interim in qualifying "low-risk" assets, and is usually located in a country like Bermuda or the Cayman Islands having a tax advantage. Insurers or other issuers of these bonds collect the bond premiums from investors in advance and receive also the returns from the investment income from these premiums. In return, investors are given a significantly higher interest rate on these cat bonds than they could receive from other investments. But some portion of both the principal and the interest on these bonds may be lost (i.e., paid to the insurer) in the event of a catastrophic event.

This is a continuously developing market as new kinds of risk are being presented to buyers and investors. Such securities covering catastrophic risks are being issued for 3-4 years. As different kinds of stocks and bonds are issued, and as these become more familiar to investors, they may become more popular and pricing of these may become more stable. In recent times, however, the price premium attached to catastrophe bonds has been very high relative to traditional reinsurance, and there has been no great rush to substitute this sort of risk-bearing capital for traditional reinsurance. But this situation could change rapidly if, as expected, reinsurance premiums begin to harden again. It seems therefore not only possible but likely that insurers and national insurance pools may find cover for catastrophic natural losses as well as earthquake risk insurance from the stock markets.

5. CONCLUSIONS

In Turkey, the major factor in our insurance sector's dependence abroad is earthquake insurance. Our insurance companies have faced significant difficulties providing cover for earthquake risks and in obtaining reinsurance cover for these risks in their portfolios. In order to reduce this dependence, taking the examples from insurance market applications in developed countries, the Decree Pool Law issued at the Official Newspaper dated 27 December 1999 makes earthquake insurance for dwellings compulsory in Turkey. To implement this, the Turkish Catastrophe Insurance Pool (TCIP) has been set up. The Government has also put into force another Decree Law concerned with better supervision and enforcement of building codes for future construction.

With the operations of the TCIP, for the time being, the losses of the victims of earthquakes will be covered, and in the future losses of victims of floods and other natural hazards will be covered and the Turkish insurance sectors' and Government's burden from catastrophic losses will be decreased. Physical damage will be covered, assuring that losses for property owners and the national economy will have the benefits of insurance, reinsurance and other risk-bearing institutions for catastrophic losses.